Michele Gianella

**Analysis of Surgical Smoke by Infrared Laser Spectroscopy**

Michele Gianella

# Analysis of Surgical Smoke by Infrared Laser Spectroscopy

## Analysis of Surgical Smoke by Infrared Laser Spectroscopy

Südwestdeutscher Verlag für Hochschulschriften

**Impressum/Imprint (nur für Deutschland/only for Germany)**
Bibliografische Information der Deutschen Nationalbibliothek: Die Deutsche Nationalbibliothek verzeichnet diese Publikation in der Deutschen Nationalbibliografie; detaillierte bibliografische Daten sind im Internet über http://dnb.d-nb.de abrufbar.
Alle in diesem Buch genannten Marken und Produktnamen unterliegen warenzeichen-, marken- oder patentrechtlichem Schutz bzw. sind Warenzeichen oder eingetragene Warenzeichen der jeweiligen Inhaber. Die Wiedergabe von Marken, Produktnamen, Gebrauchsnamen, Handelsnamen, Warenbezeichnungen u.s.w. in diesem Werk berechtigt auch ohne besondere Kennzeichnung nicht zu der Annahme, dass solche Namen im Sinne der Warenzeichen- und Markenschutzgesetzgebung als frei zu betrachten wären und daher von jedermann benutzt werden dürften.

Verlag: Südwestdeutscher Verlag für Hochschulschriften GmbH & Co. KG
Dudweiler Landstr. 99, 66123 Saarbrücken, Deutschland
Telefon +49 681 37 20 271-1, Telefax +49 681 37 20 271-0
Email: info@svh-verlag.de

Approved by: Zürich, ETH Zürich, Diss., 2011

Herstellung in Deutschland:
Schaltungsdienst Lange o.H.G., Berlin
Books on Demand GmbH, Norderstedt
Reha GmbH, Saarbrücken
Amazon Distribution GmbH, Leipzig
**ISBN: 978-3-8381-2315-8**

**Imprint (only for USA, GB)**
Bibliographic information published by the Deutsche Nationalbibliothek: The Deutsche Nationalbibliothek lists this publication in the Deutsche Nationalbibliografie; detailed bibliographic data are available in the Internet at http://dnb.d-nb.de.
Any brand names and product names mentioned in this book are subject to trademark, brand or patent protection and are trademarks or registered trademarks of their respective holders. The use of brand names, product names, common names, trade names, product descriptions etc. even without a particular marking in this works is in no way to be construed to mean that such names may be regarded as unrestricted in respect of trademark and brand protection legislation and could thus be used by anyone.

Publisher: Südwestdeutscher Verlag für Hochschulschriften GmbH & Co. KG
Dudweiler Landstr. 99, 66123 Saarbrücken, Germany
Phone +49 681 37 20 271-1, Fax +49 681 37 20 271-0
Email: info@svh-verlag.de

Printed in the U.S.A.
Printed in the U.K. by (see last page)
**ISBN: 978-3-8381-2315-8**

Copyright © 2011 by the author and Südwestdeutscher Verlag für Hochschulschriften GmbH & Co. KG and licensors
All rights reserved. Saarbrücken 2011

# Contents

| | |
|---|---|
| **Abstract** | **3** |
| **Symbols and Abbreviations** | **5** |
| Symbols | 5 |
| Abbreviations | 8 |
| **1 Introduction** | **11** |
| 1.1 Surgical Smoke | 11 |
| 1.2 Challenges | 13 |
| 1.3 Goal and Outline | 14 |
| **2 Laser Spectrometers** | **23** |
| 2.1 Infrared Laser Spectroscopy | 23 |
| 2.2 Infrared Laser Sources | 26 |
| 2.2.1 Difference Frequency Generation | 26 |
| 2.2.2 Diode and Quantum Cascade Lasers | 30 |
| 2.3 Transmission and WM Spectroscopy | 32 |
| 2.3.1 Transmission Spectroscopy | 33 |
| 2.3.2 Wavelength Modulation Spectroscopy | 35 |
| 2.3.3 Signal-to-noise Ratio with Multipass Cells | 42 |
| 2.4 DFG Based Laser Spectrometer | 47 |
| 2.4.1 Setup | 47 |
| 2.4.2 Tuning | 49 |
| 2.4.3 Data Acquisition and Evaluation | 55 |
| 2.4.4 Characterization of the DFG Spectrometer | 59 |
| 2.5 Distributed Feedback Laser Diode Spectrometer | 64 |
| 2.5.1 Setup | 64 |
| 2.5.2 Data Acquisition and Evaluation | 65 |
| 2.5.3 Calibration | 66 |
| 2.5.4 Characterization of the Laser Diode Spectrometer | 69 |

    2.6   External Cavity Quantum Cascade Laser Spectrometer . . .   73
           2.6.1   Setup . . . . . . . . . . . . . . . . . . . . . . . . . . .   73
           2.6.2   Data Acquisition and Evaluation . . . . . . . . . . . .   74
           2.6.3   Characterization of the ECQCL Laser Spectrometer .   74
           2.6.4   Preliminary Measurements . . . . . . . . . . . . . . .   77

# 3 Analysis of Infrared Spectra   79
    3.1   Spectra of Multicomponent Gas Mixtures . . . . . . . . . . . .   79
    3.2   Formulation of the Problem . . . . . . . . . . . . . . . . . . .   81
    3.3   Principal Component Analysis . . . . . . . . . . . . . . . . . .   82
    3.4   Improved Mix-Match Algorithm . . . . . . . . . . . . . . . . .   84

# 4 Measurements on *In Vitro* Samples   87
    4.1   Smoke Production and Sampling . . . . . . . . . . . . . . . .   87
    4.2   Analysis of Smoke Samples . . . . . . . . . . . . . . . . . . .   91
    4.3   Discussion . . . . . . . . . . . . . . . . . . . . . . . . . . . . .   98

# 5 Measurements on Surgical Smoke   103
    5.1   Sample Collection . . . . . . . . . . . . . . . . . . . . . . . .   103
    5.2   Results and Discussion . . . . . . . . . . . . . . . . . . . . . .   105
           5.2.1   Measurements with the DFG Spectrometer . . . . . . .   106
           5.2.2   Measurements with the FTIR Spectrometer . . . . . .   118
           5.2.3   Measurements with the DFB Spectrometer . . . . . . .   121
           5.2.4   Summary of the Obtained Results . . . . . . . . . . .   127

# 6 Conclusions and Outlook   129

# Bibliography   133

# Publications   145

# Acknowledgments   149

# Abstract

Operation room personnel and patients (particularly in laparoscopy) are exposed every day to *surgical smoke* – a by-product of the pyrolysis and combustion of biological tissue caused by heat-generating surgical equipment. Surgical smoke comprises gases, vapors, biological fragments and particulate matter. Although its *potential* danger is generally recognized and numerous studies have been carried out over the years, a final answer with respect to its health hazard has not yet been given. This is mainly due to the lack of quantitative information about its chemical composition.

In contrast to more conventional analytical techniques usually employed for the analysis of surgical smoke – such as gas chromatography-mass spectrometry – this thesis focuses for the first time on infrared laser spectroscopy. Two laser spectrometers were used: one difference frequency generation (DFG) based tunable between 2817 and 3144 $cm^{-1}$, and one featuring two distributed feedback diodes for the detection of carbon monoxide (at 2323.62 nm) and hydrogen fluoride (at 2433.13 nm). A pulsed Nd:YAG laser (1064.5 nm, 300 mW average output power, 5 kW peak power, 6 ns pulses at 4–8 kHz repetition rate) used as pump source, and a cw tunable external cavity diode laser (1520– 1600 nm, 5 mW average output power) used as signal source yield the idler beam with 150 $\mu$W average power and 150 MHz linewidth. The difference frequency generation takes place in a periodically poled lithium niobate crystal with eight poling periods and a length of 5 cm. The spectrometer is fully automated and can scan over up to 244 $cm^{-1}$ continuously without manual intervention.

Part of this work was dedicated to the adaptation, improvement and implementation of an algorithm aimed at providing the quantitative composition of a sample given its measured infrared absorption spectrum. This is done with the help of a database of infrared spectra of pure compounds. The algorithm can provide the correct constituents of a gas mixture even if there are species that cannot be identified because they do not appear in the database.

First samples were produced in the lab by cauterizing animal meat, allowing control over the amount of generated smoke, the employed animal tissue, and the atmosphere. Apart from a few relatively harmless hydrocarbons (methane, ethane and ethylene) with concentrations of up to tens of ppm, no other compound could be found with the DFG spectrometer. Additional measurements with a Fourier-transform infrared (FTIR) spectrometer revealed carbon monoxide, nitric and nitrous oxide, acetylene and hydrogen cyanide, all with concentrations around 50 ppm, except carbon monoxide for which 200 ppm were measured. No significant correlation between atmosphere, tissue and sample composition could be found.

Further measurements were carried out on actual surgical smoke collected during routine keyhole surgery at the University hospital Zurich. Methane was measured in most *in vivo* samples at concentrations of up to a few ppm. Ethane and ethylene were detected rarely, also at concentrations of a few ppm. The recommended exposure limit (REL) is 1% for all three substances. Interestingly, traces of the employed anesthetic could be found in several samples with concentrations of up to 450 ppm (REL: 2 ppm). The achieved sensitivity of the DFG spectrometer was $\alpha_{min} = 8.7 \times 10^{-7}$ cm$^{-1}$.

Carbon monoxide concentrations measured with wavelength modulation at 2323.6 nm in six samples ranged from < 250 ppb up to 3.2 ppm (REL: 30 ppm). Earlier findings of carbon monoxide concentrations in excess of 490 ppm could not be confirmed.

Finally, we tested whether the use of electrosurgical equipment in an atmosphere containing vapor of a fluorinated anesthetic (sevoflurane) could lead to the production of hydrogen fluoride (HF). Even though we did find traces of HF in samples produced in the lab, measured with wavelength modulation spectroscopy at 2433.1 nm, none of the actual surgical smoke samples contained any HF (detection limit: 110 ppt, minimum measurable absorption coefficient: $\alpha_{min} = 1.0 \times 10^{-8}$ cm$^{-1}$).

Although for the detection of some toxic compounds (e.g., benzene, formaldehyde) the sensitivity would need to be improved, many other species may readily be detected below their respective REL value. The fact that such compounds (e.g., toluene, hexane, styrene) remained undetected should put the health risk associated with them into perspective.

# Symbols and Abbreviations

## Symbols

The meaning of symbols with multiple definitions is clear from the context they appear in.

| Symbol | Description |
| ---: | --- |
| $A$ | signal area |
| $\mathscr{A}$ | absorbance |
| $\alpha$ | absorption coefficient |
| $A_\text{T}, A_\text{R}$ | transmission and reference signal area |
| $B$ | bandwidth |
| $\mathscr{B}$ | baseline |
| $\beta$ | detector noise relative to laser power |
| $c$ | concentration |
| $c$ | speed of light (vacuum) |
| $D$ | spectral database (matrix) |
| $D_\text{T}, D_\text{R}$ | transmission and reference signal |
| $\mathscr{D}$ | duty cycle |
| $\Delta$ | ADC resolution relative to full scale |
| $\Delta\phi$ | phase-mismatch (angle) |
| $\Delta k$ | phase-mismatch per unit length |
| $\Delta\lambda$ | wavelength tuning range |
| $\Delta\tilde{\nu}$ | wavenumber step size |
| $\Delta t$ | sampling point time separation |
| $\delta\mathscr{A}$ | minimum measurable absorbance |
| $\delta\alpha$ | minimum measurable absorption coefficient |

| Symbol | Description |
|---|---|
| $\delta c$ | minimum measurable concentration |
| $\delta \nu$ | frequency modulation depth |
| $\delta \nu_{\text{line}}$ | absorption linewidth (HWHM) |
| $\delta \mathscr{T}$ | minimum measurable transmittance |
| $\boldsymbol{\mathscr{E}}$ | electrical field vector |
| $\boldsymbol{\epsilon}$ | residual vector |
| $\epsilon_0$ | dielectric constant |
| $F_N$ | Nyquist frequency |
| $\mathfrak{F}_k$ | $k$th Fourier coefficient |
| $f$ | frequency |
| $\mathfrak{f}$ | modulation frequency |
| $f$ | focal length |
| $\phi_k$ | phase angle of the $k$th Fourier coefficient |
| $f_r$ | repetition rate |
| $\eta$ | conversion efficiency |
| $I$ | optical intensity |
| $\mathfrak{I}_p$ | $p \times p$ identity matrix |
| k | Boltzmann constant |
| $\kappa$ | proportionality factor |
| $k_A, k_B$ | calibration constants |
| $L$ | length of PPLN |
| $\mathscr{L}$ | absorption pathlength |
| $\Lambda$ | poling period |
| $l$ | interaction length |
| $\lambda$ | wavelength |
| $l_c$ | coherence length |
| $\lambda_s^{\text{opt}}$ | optimum signal laser wavelength |
| $m$ | modulation index |
| $m^{\text{mol}}$ | molar mass |
| $N$ | particle density |
| $\mathfrak{N}$ | number of measurements |
| $\mathscr{N}$ | noise power |
| $n$ | integer number |
| $n$ | refractive index |
| $n^*$ | optimum number of passes |
| $\nu$ | optical frequency |
| $\nu_c$ | central laser frequency |
| $\boldsymbol{P}$ | polarization vector |
| $\mathscr{P}$ | optical power |
| $p$ | number of wavelength points |

| Symbol | Description |
| --- | --- |
| $p$ | pressure |
| $Q$ | detector signal ratio |
| $R$ | gas constant |
| $R$ | SNR with integration relative to SNR of peak |
| $\mathscr{R}$ | mirror reflectivity |
| $r$ | temperature ramp slope |
| $\rho$ | mass density |
| $\rho$ | walk-off angle |
| $r_k$ | rating of $k$th substance |
| $\mathscr{S}_d$ | maximum change of incident power when scanning over an absorption line (direct transmission) |
| $\mathscr{S}_w$ | maximum change of incident power when scanning over an absorption line (wavelength modulation) |
| SNR | signal-to-noise ratio |
| $\boldsymbol{s}$ | database spectrum |
| $\boldsymbol{s}$ | sampled signal |
| $s$ | number of substances |
| $s$ | r.m.s. value of baseline drift-induced error in transmittance |
| $\sigma$ | absorption cross section |
| $\sigma$ | standard deviation (noise) |
| $s_A, s_B$ | absolute noise of normalized signal |
| $T$ | temperature |
| $\mathscr{T}$ | transmittance |
| $\mathfrak{T}$ | total acquisition time |
| $t$ | time |
| $\tau$ | signal decay time |
| $\tau$ | single scan duration |
| $\boldsymbol{U}$ | scores of $\boldsymbol{D}$ |
| $\boldsymbol{V}$ | principal components of $\boldsymbol{D}$ |
| $V$ | volume |
| $w$ | beam waist radius |
| $w$ | signal decay time (in sample points) |
| $\boldsymbol{x}$ | spectrum |
| $x$ | normalized optical frequency deviation |
| $x_c$ | normalized optical central frequency deviation |
| $\chi^{(k)}$ | $k$th order dielectric susceptibility tensor |
| $\boldsymbol{y}$ | scores of $\boldsymbol{x}$ |
| $y$ | time-dependent signal |
| $\tilde{y}$ | spectrum of $y$ |
| $y_p$ | periodic continuation of $y$ |

| Symbol | Description |
|---|---|
| $\hat{y}_k$ | $k$th Fourier coefficient of $y_p$ |
| $z$ | accelerated version of $y_p$ |

# Abbreviations

| Abbreviation | Meaning |
|---|---|
| ADC | analog-to-digital converter |
| AR | anti-reflection |
| CAS | Chemical Abstracts Service |
| CRD | cavity ring-down |
| CW | continuous wave |
| DFB | distributed feedback |
| DFG | difference frequency generation |
| DROPO | doubly resonant optical parametric oscillator |
| ECDL | external cavity laser diode |
| ECQCL | external cavity quantum cascade laser |
| FM | frequency modulation |
| FTIR | Fourier-transform infrared |
| GC | gas chromatography |
| GC-MS | gas chromatography-mass spectrometry |
| HTMC | high-temperature multipass cell |
| ICL | interband cascade laser |
| IR | infrared |
| MPC | multipass cell |
| Nd:YAG | neodymium-doped yttrium aluminum garnet |
| NICE-OHMS | noise-immune cavity enhanced optical heterodyne molecular spectroscopy |
| OPG | optical parametric generation |
| OPO | optical parametric oscillator |
| OR | operation room |
| PA | photoacoustics |
| PCA | principal component analysis |
| PM | polarization-maintaining |
| PNNL | Pacific Northwest National Laboratory |
| PPLN | periodically poled lithium niobate |

| Abbreviation | Meaning |
| --- | --- |
| PZT | piezoelectric transducer |
| QCL | quantum cascade laser |
| QPM | quasi phase-matching |
| R.M.S. | root mean square |
| REL | recommended exposure limit |
| RF | radio frequency |
| SFG | sum frequency generation |
| SHG | second harmonic generation |
| SIFT-MS | selected ion flow tube-mass spectrometry |
| SMSR | side-mode suppression ratio |
| SNR | signal-to-noise ratio |
| SROPO | singly resonant optical parametric oscillator |
| TD | thermal desorption |
| TE | thermoelectric (Peltier) |
| TROPO | triply resonant optical parametric oscillator |
| USZ | University hospital Zurich |
| WM | wavelength modulation |
| WMS | wavelength modulation spectroscopy |

# Chapter 1

# Introduction

## 1.1 Surgical Smoke

In surgery, pyrolysis and combustion of biological tissue with lasers, ultrasonic (harmonic) scalpels, high-frequency electroknives and high-speed drills/saws produce a gaseous by-product called *surgical smoke*. Such devices are used to cut, dissect, ablate, coagulate, fulgurate or vaporize biological tissue. Surgical smoke is a mixture of gases, vapors, biological and fine particulate matter.

Operation room (OR) personnel and patients are exposed to surgical smoke daily. From a health and safety perspective – for both the OR personnel and the patients – the chemical, biological and particulate compositions of surgical smoke are therefore of great interest. Surgical smoke has been shown to be a viable transport mechanism for viruses [1–5], blood and cell-containing aerosols [5–7] and tissue fragments [8]. Although rare, there have been reports of diseases transmitted from the patient to the surgeon via surgical smoke [2, 3].

A few hundred chemical species have been found so far in surgical smoke. A selection of compounds produced by cauterizing animal meat or during surgery on animals is presented in Tab. 1.1 on page 15. Most of them were detected with gas-chromatography (GC), usually coupled with mass spectrometry (GC-MS). Although GC-MS is extremely sensitive and selective, obtaining quantitative results (concentrations) requires tedious calibrations. That explains why quantitative results are only seldom given. Even when quantitative information is available, the results from different studies cannot be compared directly since there doesn't exist a standard sampling or smoke production technique.

*In vitro* studies have a few advantages over *in vivo* studies. All the parameters relevant to the production and composition of surgical smoke – setting of the surgical instrument, atmosphere, application time, tissue – can be freely changed without danger. The collection of gas samples is simple, the experiments can be repeated to test reproducibility, and the amount of pyrolyzed/cauterized tissue can be measured precisely.

Listed in Tab. 1.2 on page 20 are substances detected in surgical smoke produced during human surgery. Recent *in vivo* studies were carried out with selected ion flow tube mass spectrometry (SIFT-MS) [9] and GC/GC-MS [10–12]. The identified compounds are mainly aromatic hydrocarbons, alkenes, alkanes, aldehydes, nitriles, and ketones. Many of them are toxic and/or carcinogenic (benzene, xylene, carbon monoxide, formaldehyde, hydrogen cyanide...). The toxicity and mutagenicity of surgical smoke has been said to be at least as severe as that of cigarette smoke [11, 13]. Weston and co-workers [10] were able to detect, among others, benzene (at 5 ppb) and formaldehyde (at 5.8 ppb), two known carcinogenics [14, 15]. These concentrations are, however, well below the recommended exposure limits (REL[1]) of 500 and 300 ppb, respectively [16]. In the same study, the carbon monoxide concentration was monitored, and values in excess of 490 ppm (REL: 30 ppm) were recorded. These two examples show that estimating the health hazard linked to the exposure to surgical smoke requires knowledge of its quantitative composition.

In laparoscopic (keyhole, minimally invasive) surgery [17], vessel sealing devices that cut with minimal bleeding by applying a radio-frequency (RF) current between two electrodes are often used [18]. The abdominal cavity (*peritoneum*) is filled with an insufflant gas, typically $CO_2$ [19], within which the smoke-generating tissue pyrolysis/combustion takes place. This is in contrast to open surgery, where similar devices are used in an oxygen-rich atmosphere. Sampling smoke from laparoscopic interventions is trivial: without the need for a pump, it is sufficient to connect a sampling bag to one of the trocars [17] and let the slight overpressure of the *pneumoperitoneum* (∼20 mbar) fill the bag.

---

[1] Recommended exposure limits given by [16] are 8-hours time-weighted averages.

## 1.2 Challenges

The chemical analysis of surgical smoke presents several challenges. In the first place there is the sampling. While smoke produced *in vitro* in the lab can be sampled without concerns, hospitals have strict rules about devices brought into the OR, especially those that come close to or even in contact with the patient (electromagnetic interference from electrical devices, sterility). Ideally, one would like to work without a pump, to avoid the risk of contaminating the samples with lubricants and residues left inside it from previous samples. As pointed out in the previous section, this can be achieved easily in laparoscopy, by simply connecting a sample bag to one of the trocars via a sterile gas tube. The sample bag should be of a material that prevents molecules to adsorb to its walls, and that doesn't let species diffuse from the inside to the outside (and vice versa). Furthermore, the bag material should be chemically inert, and there should be no contamination of the sample due to desorption of chemicals from the walls (used in the manufacturing process of the bag, for example).

In the previous section it was shown that the composition of surgical smoke is extremely manifold, and that low concentrations (below parts-per-million) should be expected for several components. This requires a selective and sensitive detection technique. Moreover, quantitative results should be easily obtainable, as health risk assessments due to exposure to toxic gases can only be made if concentrations (and exposure times) are known. A few analytical techniques have already been mentioned in the previous section (GC, GC-MS, SIFT-MS). Infrared absorption spectroscopy has not been mentioned, due to the almost complete lack of studies involving it. In infrared (IR) spectroscopy, the vibrational-rotational structure of molecules in the sample is probed, providing quantitative and qualitative information [20, 21]. The Fourier-transform infrared (FTIR) spectrometer [22] is without doubt the workhorse of IR spectroscopy, but for gas detection at trace levels laser-based spectrometers are often used [23]. Their inherent high spectral resolution makes them extremely selective, while high power spectral density and good spatial coherence allow sensitive detection schemes with long absorption pathlengths [24–31].

## 1.3 Goal and Outline

The goal of this study was to investigate the chemical composition of surgical smoke produced during laparoscopic surgery with laser-spectroscopic techniques. In Chapter 2 we present the laser spectrometers used in this study and their key features. In Chapter 3 we describe an algorithm that we developed which can be used in conjunction with a spectral library to identify and quantify components of a gas mixture based on its absorption spectrum. In Chapters 4 and 5 we present and discuss results obtained with surgical smoke samples produced *in vitro* and *in vivo*. Finally, in Ch. 6, some concluding remarks and outlook are given.

**Table 1.1** – Selection of chemical compounds detected in *surgical* smoke produced with animal meat *in vitro*.

| Name | CAS Nr. | Refs. |
| --- | --- | --- |
| Pyrrole | 109-97-7 | [32–42] |
| Methylpyrazine | 109-08-0 | [32, 33, 35–39, 41, 43] |
| Benzaldehyde | 100-52-7 | [33, 36–38, 40–42, 44] |
| Isovaleraldehyde | 590-86-3 | [34–38, 40, 43] |
| Benzonitrile | 100-47-0 | [32, 33, 36, 37, 41, 44] |
| Phenylacetylene | 536-74-3 | [32, 33, 36, 37, 41, 44] |
| 2-Methylbutanenitrile | 18936-17-9 | [32, 35–37, 39] |
| 3-Methylbutanenitrile | 625-28-5 | [32, 35–37, 39] |
| Pyridine | 110-86-1 | [32, 36, 37, 40–42] |
| 2-Methylbutanal | 96-17-3 | [36–38, 43] |
| 4-Methylphenol | 106-44-5 | [32, 33, 37, 41] |
| Phenol | 108-95-2 | [32, 37, 41, 44] |
| 1-Methylpyrrole | 96-54-8 | [36, 37, 39, 42] |
| 2,5-Dimethylfurane | 625-86-5 | [35–37] |
| 2,5-Dimethylpyrazine | 123-32-0 | [32, 35, 38] |
| 2-Methylpropanal | 78-84-2 | [37, 38, 43] |
| 3-Butenenitrile | 109-75-1 | [33, 37, 41] |
| Benzenepropanenitrile | 645-59-0 | [32, 36, 37] |
| Butyraldehyde | 123-72-8 | [37, 38, 40] |
| Dimethyldisulfide | 624-92-0 | [32, 36, 37] |
| Furfurylalcohol | 98-00-0 | [32, 36, 37] |
| Methylisothiocyanate | 556-61-6 | [32, 36, 37] |
| Naphthalene | 91-20-3 | [32, 36, 44] |
| Phenylacetonitrile | 140-29-4 | [32, 36, 37] |
| Indane | 496-11-7 | [32, 33, 41] |
| Palmitic acid | 57-10-3 | [32, 33, 41] |
| Indole | 120-72-9 | [32, 33, 41] |
| 1-Ethylpyrrole | 617-92-5 | [36, 37] |
| 1-Octene | 111-66-0 | [36, 37] |
| 1-Propenylbenzene | 637-50-3 | [32, 37] |
| 1-Tridecene | 2437-56-1 | [36, 37] |
| 2,6-Dimethylpyrazine | 108-50-9 | [32, 37] |
| 2-Ethylpyridine | 100-71-0 | [32, 37] |
| 2-Methylpropanol | 78-83-1 | [33, 41] |
| 2-Methylfurane | 534-22-5 | [35, 37] |
| 2-Methylpropanenitrile | 78-82-0 | [37, 38] |

**Table 1.1** – *(continued)* Selection of chemical compounds detected in *surgical* smoke produced with animal meat *in vitro*.

| Name | CAS Nr. | Refs. |
| --- | --- | --- |
| 2-Methylpropylbenzene | 538-93-2 | [32, 37] |
| 4-Methylvaleronitrile | 542-54-1 | [35, 37] |
| 5-Methylfurfural | 620-02-0 | [32, 37] |
| 6-Methylindole | 3420-02-8 | [33, 41] |
| Acetaldehyde | 75-07-0 | [35, 38] |
| Acetonitrile | 75-05-8 | [41, 43] |
| Dimethyltrisulfide | 3658-80-8 | [32, 37] |
| Indene | 95-13-6 | [32, 36] |
| Pyrazine | 290-37-9 | [33, 37] |
| Trimethylpyrazine | 14667-55-1 | [33, 37] |
| 2-Butanone | 78-93-3 | [38, 40] |
| 1-Nonene | 124-11-8 | [37] |
| Acetic acid | 64-19-7 | [32] |
| Decanol | 112-30-1 | [32] |
| Dodecanol | 112-53-8 | [32] |
| Ethane | 74-84-0 | [41] |
| Heptane | 142-82-5 | [36] |
| Methane | 74-82-8 | [41] |
| 1-Methylnaphthalene | 90-12-0 | [32] |
| 1-Pentadecene | 13360-61-7 | [37] |
| 2,3-Dimethylpyrazine | 5910-89-4 | [32] |
| 2,3-Dimethylpyridine | 583-61-9 | [32] |
| 2,4-Dimethylphenol | 105-67-9 | [32] |
| 2,4-Dimethylpyridine | 108-47-4 | [32] |
| 2,5-Dimethylpyridine | 589-93-5 | [32] |
| 2-Acetyl-5-methylfuran | 1193-79-9 | [32] |
| 2-Acetylfuran | 1192-62-7 | [32] |
| 2-Acetylthiophene | 88-15-3 | [32] |
| 2-Aminobenzonitrile | 1885-29-6 | [32] |
| 2-Cyanopyridine | 100-70-9 | [32] |
| 2-Ethyl-2-butenal | 19780-25-7 | [32] |
| 2-Ethyl-3,6-dimethylpyrazine | 13360-65-1 | [32] |
| 2-Ethyl-5-methylpyrazine | 13360-64-0 | [32] |
| 2-Ethylbutyraldehyde | 97-96-1 | [35] |
| 2-Formylthiophene | 98-03-3 | [32] |
| 2-Methyl-2-butenal | 1115-11-3 | [32] |

**Table 1.1** – *(continued)* Selection of chemical compounds detected in *surgical* smoke produced with animal meat *in vitro*.

| Name | CAS Nr. | Refs. |
| --- | --- | --- |
| 2-Methyl-2-pentenal | 623-36-9 | [32] |
| 2-Methyl-2-Propenal | 78-85-3 | [37] |
| 2-Methyl-5-vinylpyrazine | 13925-08-1 | [32] |
| 2-Methyl-6-vinylpyrazine | 13925-09-2 | [32] |
| 2-Methylaminopyridine | 4597-87-9 | [32] |
| 2-Methylnaphthalene | 91-57-6 | [32] |
| 2-Methylphenol | 95-48-7 | [32] |
| 2-Methylpyridine | 109-06-8 | [32] |
| 2-Methylquinoxaline | 7251-61-8 | [32] |
| 2-Methylstyrene | 611-15-4 | [32] |
| 2-Phenyl-2-butenal | 4411-89-6 | [32] |
| Benzeneethanol | 60-12-8 | [32] |
| 2-Propionylpyridine | 3238-55-9 | [32] |
| 2-Vinylpyridine | 100-69-6 | [32] |
| 3-(N,N-Dimethylamino)-propionitrile | 1738-25-6 | [32] |
| 3-Butenylbenzene | 768-56-9 | [37] |
| 3-Formylthiophene | 498-62-4 | [32] |
| 3-Methylpyridine | 108-99-6 | [32] |
| 3-Methylstyrene | 100-80-1 | [32] |
| 3-Phenylpyridine | 1008-88-4 | [32] |
| 4-Ethylphenol | 123-07-9 | [32] |
| 4-Methyl-2-Pentanon | 108-10-1 | [36] |
| 4-Methylpentanoic acid | 646-07-1 | [32] |
| 4-Methylpyridine | 108-89-4 | [32] |
| 4-Phenylbutan-2-one | 2550-26-7 | [32] |
| 4-Phenylbutyronitrile | 2046-18-6 | [32] |
| 5,6,7,8-Tetrahydroquinoxaline | 34413-35-9 | [32] |
| 5-Methylfurfurylalcohol | 3857-25-8 | [32] |
| 5-Methylhexan-2-one | 110-12-3 | [32] |
| Acenaphthene | 83-32-9 | [32] |
| Acetophenone | 98-86-2 | [32] |
| Acroloin | 107-02-8 | [41] |
| Benzylmethyldisulfide | 699-10-5 | [32] |
| Benzylmethylsulfide | 766-92-7 | [32] |
| Benzylalcohol | 100-51-6 | [32] |
| Butanenitrile | 109-74-0 | [35] |

**Table 1.1** – *(continued)* Selection of chemical compounds detected in *surgical* smoke produced with animal meat *in vitro*.

| Name | CAS Nr. | Refs. |
| --- | --- | --- |
| Butyrophenone | 495-40-9 | [32] |
| Carbon disulphide | 75-15-0 | [45] |
| Cholesterol | 57-88-5 | [32] |
| Creosol | 93-51-6 | [41] |
| Dibenzyl | 103-29-7 | [32] |
| Diphenylmethane | 101-81-5 | [32] |
| Ethylhexadecanoate | 628-97-7 | [32] |
| Ethyloctadecanoate | 111-61-5 | [32] |
| Ethylpyrazine | 13925-00-3 | [32] |
| Furfurylmethylsulfide | 1438-91-1 | [32] |
| 2-Heptanone | 110-43-0 | [32] |
| Heptanoic acid | 111-14-8 | [32] |
| Hexadecanal | 629-80-1 | [32] |
| Hexadecane | 544-76-3 | [32] |
| 1-Hexadecanol | 36653-82-4 | [32] |
| Hexanal | 66-25-1 | [32] |
| Hexanoic acid | 142-62-1 | [32] |
| Isoquinoline | 119-65-3 | [32] |
| Limonene | 138-86-3 | [32] |
| Linoleic acid | 60-33-3 | [32] |
| Methylhexadecanoate | 112-39-0 | [32] |
| Methyloctadecanoate | 112-61-8 | [32] |
| m-Tolunitrile | 620-22-4 | [32] |
| N,N-Dimethylacetonitrile | 926-64-7 | [36] |
| N-Ethylindol | 10604-59-8 | [32] |
| n-Valeronitrile | 110-59-8 | [37] |
| Octadecanal | 638-66-4 | [32] |
| Octadecanoic acid | 57-11-4 | [32] |
| Octanal | 124-13-0 | [32] |
| 1-Octanol | 111-87-5 | [32] |
| Nonanenitrile | 2243-27-8 | [35] |
| Oleic acid | 2027-47-6 | [32] |
| Oleic acid amide | 301-02-0 | [32] |
| Pentanoic acid | 109-52-4 | [36] |
| Phenanthrene | 85-01-8 | [32] |
| Phenylacetaldehyde | 122-78-1 | [32] |

**Table 1.1** – *(continued)* Selection of chemical compounds detected in *surgical* smoke produced with animal meat *in vitro*.

| Name | CAS Nr. | Refs. |
| --- | --- | --- |
| p-Tolunitrile | 104-85-8 | [32] |
| Quinoline | 91-22-5 | [32] |
| Quinoxaline | 91-19-0 | [32] |
| Skatol | 83-34-1 | [32] |
| Tetradecanal | 124-25-4 | [32] |
| Tetradecanol | 112-72-1 | [32] |
| Tetrahydrofuran | 109-99-9 | [37] |
| Tetramethylpyrazine | 1124-11-4 | [32] |
| Trimethylacetonitrile | 630-18-2 | [35] |
| Valeraldehyde | 110-62-3 | [35] |
| Vinylpyrazine | 4177-16-6 | [32] |
| Methylacrylate | 96-33-3 | [38] |
| Methanol | 67-56-1 | [46] |
| Ethanol | 64-17-5 | [46] |
| Ammonia | 7664-41-7 | [46] |

**Table 1.2** – Selection of chemical compounds detected in surgical smoke produced during surgery.

| Name | CAS Nr. | Refs. |
| --- | --- | --- |
| Toluene | 108-88-3 | [10, 11, 32–45, 47] |
| Ethylbenzene | 100-41-4 | [10, 11, 32–42, 44, 45] |
| Styrene | 100-42-5 | [10, 35–38, 40–45] |
| Benzene | 71-43-2 | [10, 36–39, 41, 43–45] |
| Acrylonitrile | 107-13-1 | [10, 12, 34, 35, 38, 41, 43] |
| 1-Undecene | 821-95-4 | [11, 32, 33, 36, 37, 41] |
| 1-Decene | 872-05-9 | [11, 33, 36, 37, 41, 47] |
| p-Xylene | 106-42-3 | [10, 11, 32, 36, 37] |
| m-Xylene | 108-38-3 | [10, 11, 32, 36, 37] |
| Furfural | 98-01-1 | [32, 33, 37, 41, 47] |
| Propanenitrile | 107-12-0 | [34, 35, 43, 47] |
| Hydrogen cyanide | 74-90-8 | [9, 41, 44, 48] |
| Propylbenzene | 103-65-1 | [11, 32, 36, 37] |
| Isobutene | 115-11-7 | [12, 41, 43, 47] |
| 1-Heptene | 592-76-7 | [10, 36, 37, 47] |
| 1,3-Butadiene | 106-99-0 | [9, 12, 41, 47] |
| Carbon monoxide | 630-08-0 | [10, 41, 44, 48] |
| Propene | 115-07-1 | [10, 12, 41, 43] |
| Ethylene | 74-85-1 | [41, 43, 47] |
| Methylthiocyanate | 556-64-9 | [36, 37, 47] |
| 1-Dodecene | 112-41-4 | [11, 36, 37] |
| 1-Tetradecene | 1120-36-1 | [11, 36, 37] |
| Acetylene | 74-86-2 | [9, 41, 43] |
| o-Xylene | 95-47-6 | [10, 11, 36] |
| 3-Methylstyrene | 100-80-1 | [32, 47] |
| 1-Dodecane | 112-40-3 | [11, 32] |
| 1-Undecane | 1120-21-4 | [11, 32] |
| Pentadecane | 629-62-9 | [11, 36] |
| Tetradecane | 629-59-4 | [11, 32] |
| Decane | 124-18-5 | [11, 32] |
| Formaldehyde | 50-00-0 | [10, 41] |
| 1-Butene | 106-98-9 | [10, 41] |
| Acetone | 67-64-1 | [10, 37] |
| 1-Pentene | 109-67-1 | [10, 12] |
| Heptanal | 111-71-7 | [11] |
| Nonanal | 124-19-6 | [11] |

**Table 1.2** – *(continued)* Selection of chemical compounds detected in surgical smoke produced during surgery.

| Name | CAS Nr. | Refs. |
|---|---|---|
| Cyclohexanone | 108-94-1 | [11] |
| Perchloroethylene | 127-18-4 | [11] |
| Tridecane | 629-50-5 | [11] |
| Ammonia | 7664-41-7 | [47] |
| 1-Hexene | 592-41-6 | [10] |
| Isooctane | 540-84-1 | [10] |
| Propadiene | 463-49-0 | [12] |
| Vinylacetylene | 689-97-4 | [12] |
| Mercaptomethane | 74-93-1 | [12] |
| Ethylacetylene | 107-00-6 | [12] |
| Diacetylene | 460-12-8 | [12] |
| Ethanol | 64-17-5 | [12] |
| Piperylene | 540-60-9 | [12] |
| Propenylacetylene | 2206-23-7 | [12] |
| 1,4-Pentadiene | 591-93-5 | [12] |
| Cyclopentadiene | 542-92-7 | [12] |
| Butyrolactone | 96-48-0 | [12] |

# Chapter 2

# Laser Spectrometers

## 2.1 Infrared Laser Spectroscopy

The energy of photons in the infrared (IR) part of the electromagnetic spectrum (0.7– 1000 µm) corresponds to transitions between vibrational and rotational states of molecules. With IR spectroscopy one can therefore probe the vibrational structure of molecules in a gas, liquid or solid sample. Vibrational modes can be classified into *stretching* (bond lengths change) and *bending* (bond angles change) modes. These give rise to IR absorption bands or lines. The positions and intensities thereof depend on the vibrational modes involved and of the remainder of the molecule attached to the vibrating part. The intensity of an absorption is also linked to the abundance of the molecule. A skilled spectroscopist can make a qualitative statement about the composition of a sample by looking at its spectrum and finding the positions and intensities of absorption bands. For example, alkanes (hydrocarbons with only single bonds) manifest absorption bands at 2850–3000 cm$^{-1}$ (strong, C–H stretch modes) and around 1375, 1450, and 1465 cm$^{-1}$ (bending modes) [21]. Matching a definite substance to a measured spectrum and obtaining quantitative information, however, requires a library (database) of IR spectra to be used for comparison. For mixtures with relatively few components this work can be done "by hand", by visually comparing the measured spectrum with spectra from the database, and then fitting the spectra of the identified substances to the measured spectrum. For mixtures with many components this is no longer possible and needs to be automated (see Ch. 3).

While absorptions can be found all over the IR range, there are a few so-called *fingerprint* regions that are particularly interesting for spectroscopy.

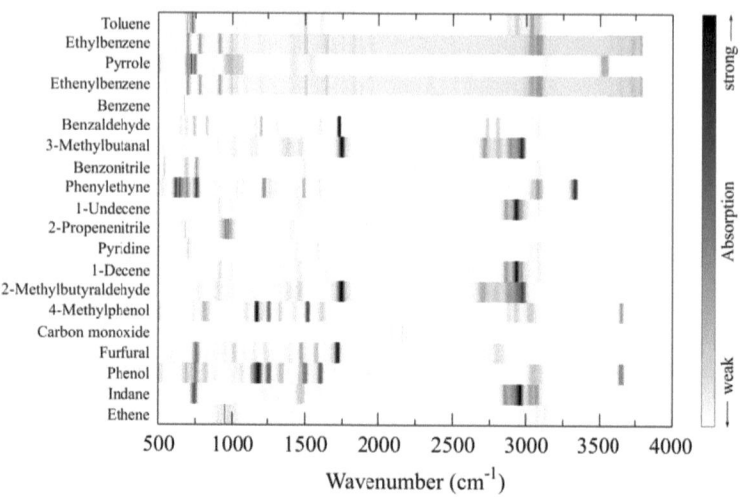

**Figure 2.1** – Infrared absorption spectra of 20 substances selected from Tabs. 1.1 and 1.2 (black = strong absorption, white = no absorption).

One of them is situated near 3000 cm$^{-1}$ (3.3 μm). There, molecules with C–H bonds have relatively strong absorption lines or bands. Figure 2.1 shows the IR absorption spectra of 20 substances selected from Tabs. 1.1 and 1.2 from 500 cm$^{-1}$ (20 μm) to 4000 cm$^{-1}$ (2.5 μm). As most of them are hydrocarbons, they all manifest absorption lines or bands near 3000 cm$^{-1}$. Another region of interest is between 500 and 2000 cm$^{-1}$ (5–20 μm). The absorption bands of a molecule (both position and intensity) depend on the chemical groups it consists of. For instance, aldehydes (benzaldehyde, 3-methylbutanal, 2-methylbutyraldehyde and furfural) have strong absorptions near 1700 cm$^{-1}$, while aromatic compounds (toluene, ethylbenzene, ethenylbenzene, benzonitrile, phenylethyne) absorb between 500 and 1000 cm$^{-1}$.

Located at shorter wavelengths (< 2.5 μm) are overtones of the stronger mid-IR absorptions. These can be several orders of magnitude weaker than their mid-IR counterparts. However, excellent readily available laser sources and detectors compensate for the smaller absorption cross-section and make this a spectroscopically interesting range.

A laser spectrometer consists of three interrelated parts: a laser source, a detection scheme and a detector. The detection scheme generates a measurable signal that depends on the sample's absorption. Some well-known schemes include direct transmission, wavelength and frequency modulation [49], cavity ring-down and leak-out [50], and photoacoustics [51]. The detector must obviously be matched to both the detection scheme and to the laser source. Frequency modulation at several MHz, for example, prevents the use of slow detectors such as thermopiles, and low-noise, sensitive Si-photodiodes can only be used with wavelengths below 1.1 $\mu$m. The laser source determines which detection schemes are possible. For example, high-power gas lasers are not suited for wavelength modulation but are excellent sources for photoacoustics, while low-power diode lasers are often used for wavelength modulation but seldom for photoacoustics.

The choice of laser source, detection scheme and detector requires careful examination of the experimental requirements. For example, some lasers and detectors require cryogenic cooling. This may be acceptable in the lab, but is generally unsuitable for unattended field measurements. There may also be constraints with respect to weight, volume and power consumption. The required sensitivity and accuracy need to be taken into account, too. The ultimate sensitivity of a spectrometer is reached when the signal produced by the sample has the same magnitude as the spectrometer's noise, which includes quantum noise (from the laser), electronic noise (thermal and flicker noise, both from laser and detector) [52], noise introduced via mechanical vibrations, ground loops and other sources of interference, and quantization noise [53]. To enhance the signal produced by very weak absorptions, long path absorption cells where the laser beam experiences multiple reflections (*multipass* cells) [54] are often used in trace gas detection. If there are potentially interfering species in the measured sample, the spectrometer needs to be selective. Selectivity is achieved through high resolution and an appropriate choice of wavelength. Similar overlapping spectra may be distinguished by reducing the pressure of the gas, thereby reducing absorption linewidths and/or resolving line structures in absorption bands. The linewidth of the laser determines the resolution of the spectrometer, with typical values below 1 MHz for external cavity diode lasers. For comparison, absorption linewidths at atmospheric pressure are of the order of several GHz [55]. The selected wavelength interval should manifest characteristic absorption features for the species under examination. The broader the tuning range of the laser source, the more substances can be detected.

## 2.2 Infrared Laser Sources

During this study a difference frequency generation (DFG) based laser spectrometer, and a spectrometer with two distributed feedback (DFB) laser diodes were used. A few measurements were performed with an external cavity quantum cascade laser (ECQCL). A brief overview of DFG and DFB lasers is given in this section.

### 2.2.1 Difference Frequency Generation

Laser sources suitable for spectroscopy are unfortunately not readily available at any wavelength. With the high intensity achievable with lasers, nonlinear optical effects can be induced in dielectrics to produce coherent radiation at new wavelengths. The lowest order nonlinear effects are second order (three-photon) processes: difference and sum frequency generation (DFG, SFG), frequency doubling (or second harmonic generation, SHG), optical parametric generation and oscillation (OPG, OPO), and optical rectification [56]. Unlike higher order nonlinear processes which can take place in any material, second order effects only happen within crystals without an inversion symmetry [57]. At high optical intensities, the relationship between the electrical field and the polarization vector of the crystal is no longer linear. The polarization vector can be expanded in a Taylor series of the electrical field $\mathscr{E}$:

$$P_i = \epsilon_0 \left( \chi_{ij}^{(1)} \mathscr{E}_j + \chi_{ijk}^{(2)} \mathscr{E}_j \mathscr{E}_k + \chi_{ijkl}^{(3)} \mathscr{E}_j \mathscr{E}_k \mathscr{E}_l + \cdots \right) \quad i = 1,2,3, \tag{2.1}$$

where $\boldsymbol{P} \equiv (P_1, P_2, P_3)$ is the polarization vector, $\boldsymbol{\mathscr{E}} \equiv (\mathscr{E}_1, \mathscr{E}_2, \mathscr{E}_3)$ is the electrical field vector, $\chi^{(k)}$ are the dielectric susceptibility tensors, $\epsilon_0$ is the dielectric constant and we have used Einstein's summation convention. The effects mentioned above are caused by the second order nonlinear susceptibility tensor $\chi_{ijk}^{(2)}$. Of interest for infrared spectroscopy are the two down-conversion processes DFG and OPG/OPO.

For DFG and OPG/OPO, mature and well-proven near infrared lasers (Nd:YAG lasers, external cavity diode lasers, distributed feedback laser diodes) can be used to generate mid infrared radiation at wavelengths where laser sources are unavailable or do not fulfill the application-specific requirements – such as continuous-wave operation, broad continuous tunability, room-temperature operation, narrow linewidth, robustness. In OPG a *pump* laser is employed to generate a *signal* and *idler* beam in a nonlinear crystal. In DFG both a pump and signal laser are used to generate an idler beam (Fig. 2.2). The signal laser is amplified, a process known as optical parametric amplification. In both OPG and DFG energy conservation is

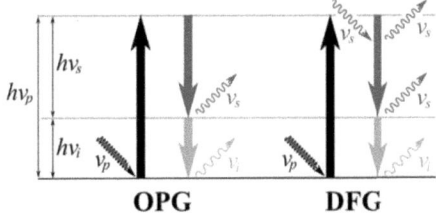

**Figure 2.2** – Schematic illustration of optical parametric generation (OPG) and difference frequency generation (DFG). Dashed lines are virtual states, straight arrows represent transitions, wavy arrows are photons ($p$: pump, $s$: signal, $i$: idler). In DFG the emission of the signal photon is stimulated by a second incident signal photon; the signal beam is therefore amplified.

fulfilled:
$$\nu_p = \nu_s + \nu_i, \qquad (2.2)$$

where $\nu_p, \nu_s$ and $\nu_i$ are the pump, signal and idler frequencies, respectively. By convention, the pump beam has the highest frequency and the idler the lowest one. In DFG both $\nu_p$ and $\nu_s$ are given and the idler frequency $\nu_i$ follows from Eq. (2.2). The idler beam inherits its characteristics – such as beam profile, spectral linewidth and coherence – from the generating beams. In particular, the lineshape of the idler is the convolution of the lineshape of pump and signal beam. The situation is different for OPG, where only $\nu_p$ is predefined and Eq. (2.2) has infinitely many solutions. The lineshape of both idler and signal beam is then determined by *phase-matching*. For many applications the linewidth of an OPG-based source is too large. If a narrower emission spectrum is desired, the nonlinear crystal can be embedded in a cavity resonant for one (singly resonant optical parametric oscillator, SROPO), two (DROPO) or all three (TROPO) wavelengths [58, 59]. The strength of OPOs lies in the very broad tuning range and in the typically high output powers [60]. But automated mode hop free tuning over more than a few cm$^{-1}$ – as is desirable for the present study – is problematic [61], as the length of the cavity and possible etalons have to be adjusted/turned synchronously. We will therefore focus our discussion on DFG from this point on.

An idler photon generated in one point of the crystal must have, once it reaches a point where another photon is being generated, the same phase as that photon, or destructive interference will take place. The condition for constructive interference is expressed as [57]

$$n_p \nu_p - n_s \nu_s - n_i \nu_i = 0, \qquad (2.3)$$

where $n_p \equiv n(\nu_p)$, $n_s \equiv n(\nu_s)$ and $n_i \equiv n(\nu_i)$ are the refractive indices at $\nu_p, \nu_s$ and $\nu_i$, respectively. Equation (2.3) is called the phase-matching condition. If Eq. (2.3) is fulfilled, the largest possible idler power is obtained. Otherwise the power is reduced because (partially) destructive interference takes place. With the phase-mismatch $\Delta k$,

$$\Delta k \equiv \frac{2\pi}{c}(n_p \nu_p - n_s \nu_s - n_i \nu_i), \tag{2.4}$$

the achieved idler power as a function of phase-mismatch is given by [62]

$$\mathcal{P}_i(\Delta k) = \eta \mathcal{P}_p \operatorname{sinc}^2(\Delta k L/2), \tag{2.5}$$

where $L$ is the length of the crystal, $\mathcal{P}_p$ is the pump power and $\eta$ is the conversion efficiency for perfect phase-matching ($\Delta k = 0$). For DFG $\eta$ is

$$\eta = \frac{2\pi \nu_i^2 \chi_{\text{eff}}^2 L^2 I_s}{\epsilon_0 c^3 n_p n_s n_i}, \tag{2.6}$$

where $\chi_{\text{eff}}$ is the effective second order susceptibility, which depends on the polarizations of the three beams with respect to the crystal axes, and $I_s$ is the intensity of the signal beam. Longer crystals provide more interaction length, but the intensities of the beams become smaller due to defocussing: if this is taken into account, the conversion efficiency is proportional only to $L$ instead of $L^2$.

Tuning a DFG source requires changing the signal or pump frequency while at the same time ensuring that the phase-matching condition Eq. (2.3) remains fulfilled. For this, one can take advantage of the temperature dependence of the refractive index (temperature tuning) or, in birefringent crystals, one can use the dependency of the refractive index on the angle between the polarization plane and the crystal axis (angle tuning). A third possibility is given by *quasi phase-matching* (QPM) [63] with *periodically poled* nonlinear crystals (Fig. 2.3). For some materials, one of the elements of $\chi_{ijk}^{(2)}$ [Eq. (2.1)] is much larger than the others. For lithium niobate (LiNbO$_3$), for example, the largest element is $\chi_{333}^{(2)} = -27$ pm/V [64]. To achieve maximal conversion efficiency $\eta$ [Eq. (2.6)], the polarizations of both pump and signal beam must therefore be along the $z$ axis of the crystal (the idler is also polarized along the $z$ axis). In this configuration, it is not possible to achieve perfect phase-matching ($\Delta k = 0$). With $\Delta k \neq 0$ a coherence length $\ell_c \equiv \pi/\Delta k$ [57] can be defined: light generated within one coherence length will interfere destructively with light generated in the previous coherence length. As a result, the idler intensity cannot build up. The order of magnitude of $\ell_c$ is typically 1–10 μm. If the domain orientation of the crystal is reversed every $\ell_c$, the phase difference $\Delta \phi$ accumulated in the first

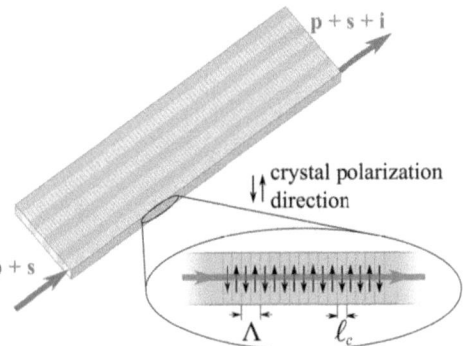

**Figure 2.3** – Periodically poled crystal with eight different poling periods Λ. The domain orientation is reversed every length $\ell_c = \Lambda/2$. Pump (p) and signal (s) beam overlap along the entire length of the crystal, and generate the idler (i).

domain, $\Delta\phi = \Delta k \ell_c = \pi$, is reset to zero when the beams enter the following domain and idler power can build up. The period of the poling is therefore $\Lambda = 2\ell_c$, and we can rewrite Eq. (2.4) as

$$\frac{1}{\Lambda} = \frac{1}{c}(n_p \nu_p - n_s \nu_s - n_i \nu_i). \tag{2.7}$$

In analogy to Eq. (2.3), Eq. (2.7) is called the quasi phase-matching condition. Compared to perfect phase-matching, the conversion efficiency of quasi phase-matched DFG is reduced by a factor $(2/\pi)^2$. For many non-linear materials this loss is overcompensated by the fact that the highest element of $\chi^{(2)}_{ijk}$ can be used. A welcome side-effect is that, since $\chi^{(2)}_{333}$ is the largest tensor element for many materials [64, 65], the polarizations of all beams are the same and spatial walk-off [1] does not occur.

The periodic domain orientation reversal that characterizes periodically poled crystals can be obtained in a few ways [66–68], the most common of which is ferroelectric domain engineering [66]. In this case, a strong electric field is applied onto a ferroelectric crystal with an arrangement of electrodes that corresponds to the desired poling structure and period. One of the most commonly used crystals is periodically poled lithium niobate (PPLN).

---

[1] Spatial walk-off limits the interaction length to $l \equiv w\sqrt{\pi}/\rho \approx 5$ mm for a beam waist $w$ of 100 μm and a walk-off angle $\rho$ of 2° (LiNbO$_3$).

## 2.2.2 Diode and Quantum Cascade Lasers

Since the advent of the quantum cascade laser (QCL) [69], lead-salt diode lasers have lost their appeal due to disadvantageous tuning characteristics and the need for cryogenic cooling [70]. QCLs can be tailored to emit at (almost) any wavelength between 3.1 and 16 $\mu$m [71, 72], whereas laser diodes based on GaAs [73], InP [74] and GaSb [75] cover the spectrum from 760 nm up to 3 $\mu$m. Recently, single-mode continuous wave (cw) emission at 3.4 $\mu$m was achieved near room temperature with a distributed feedback (DFB) quantum well laser diode based on type-I GaInAsSb/AlGaInAsSb [76]. A third type of semiconductor laser is the interband cascade laser (ICL) [77]. Unlike in QCLs, where optical emission is caused by electronic transitions within the conduction band only, in ICLs laser emission occurs in a cascade of quantum wells where electron-hole recombinations between the conduction and valence band take place [78]. ICLs are becoming increasingly interesting laser sources for spectroscopy due to their favorable emission range (3–5 $\mu$m) and room-temperature continuous wave (cw) operation [79]. Single-mode operation required for high-resolution spectroscopy can be achieved by implementing an external cavity (ECQCL/ECDL) [80] or with a DFB structure [81].

Laser sources equipped with external cavity emit on a single mode and have large tuning ranges $\Delta\lambda$ from about $\Delta\lambda/\lambda \sim 0.1$ [29, 82] up to $\Delta\lambda/\lambda = 0.39$ [83, 84]. For mode hop free tuning operation, the mirror (or diffraction grating) rotation angle and the cavity length must be tuned simultaneously [80, 85]. The external cavity, however, comes at a price of higher complexity, weight and dimensions, as well as susceptibility to mechanical vibrations.

In DFB laser diodes and DFB-QCLs, single-mode operation is achieved with a grating grown along the waveguide [81]. In contrast to ECDLs/ECQCLs, there are no movable parts and the laser is only a few mm long. This makes DFB lasers extremely robust, maintenance-free, small and light. Tuning is achieved by changing the temperature – and thereby the refractive index – of the device, either by influencing the temperature directly, or by altering the injection current. Because the injection current can be controlled and modulated much faster and more precisely than the temperature, the latter is usually set to a value appropriate for the desired emission wavelength, and the tuning occurs by modulating the injection current. Because of the tuning simplicity, DFB lasers are well suited for wavelength or frequency modulation (WM, FM) based detection schemes. Typical tuning rates are 10 GHz/K for temperature tuning and 1 GHz/mA for injection current tuning [86]. The strong dependence of emission wavelength on diode temperature requires careful thermal management. For example, at 10 GHz/K a frequency stability of 10 MHz requires a temperature stabilization of the order of 1 mK. DFB lasers have narrow linewidths (< 3 MHz), good side-mode suppression ratios (SMSR~ −55...−30 dB) and output powers in the mW up to a few hundred mW range [81]. Tuning ranges are usually a few nm wide. Such small intervals limit the number of compounds that can be detected with a single device.

## 2.3 Transmission and Wavelength Modulation Spectroscopy

Direct transmission spectroscopy was chosen in combination with the DFG spectrometer because the achieved optical power was not sufficient for more sensitive detection schemes, such as cavity ring-down or photoacoustic spectroscopy.

In both direct transmission and wavelength modulation spectroscopy, the *transmittance* (or a related quantity) of a sample is measured.

The optical power $\mathscr{P}$ of a light beam with wavelength $\lambda$ transmitted through a cell of length $\mathscr{L}$ filled with an absorbing species is given by Beer-Lambert's law:

$$\mathscr{P} = \mathscr{P}_0 \exp(-\sigma(\lambda) N \mathscr{L}), \tag{2.8}$$

where $\sigma(\lambda)$ is the pressure- and temperature-dependent absorption cross section, $\mathscr{P}_0$ is the incident optical power and $N$ is the the particle density of the absorber. With the definition of the concentration,

$$c \equiv \frac{N}{N_0}, \tag{2.9}$$

where $N_0$ is the total particle density, we can rewrite Eq. (2.8) as

$$\mathscr{P} = \mathscr{P}_0 \exp(-\sigma(\lambda) c \mathscr{L} N_0). \tag{2.10}$$

For an ideal gas, $N_0$ depends on the total pressure $p$ and on the temperature $T$ of the sample:

$$N_0 = \frac{p}{kT} \approx 2.505 \times 10^{25} \, \text{m}^{-3}, \tag{2.11}$$

at $p = 101325$ Pa and $T = 293$ K ($k = 1.381 \times 10^{-23}$ J/K is Boltzmann's constant). The transmittance $\mathscr{T}$ of the sample is defined as

$$\mathscr{T} \equiv \frac{\mathscr{P}}{\mathscr{P}_0} \tag{2.12}$$

and its absorbance $\mathscr{A}$ as

$$\mathscr{A} \equiv -\log_{10} \mathscr{T} = c \mathscr{L} \frac{\sigma(\lambda) N_0}{\ln 10}. \tag{2.13}$$

The absorbance $\mathscr{A}$ is proportional to both the pathlength $\mathscr{L}$ and the concentration $c$. The absorption coefficient $\alpha$ is defined via

$$\mathscr{T} \equiv 10^{-\alpha \mathscr{L}} \tag{2.14}$$

as
$$\alpha = \frac{\mathscr{A}}{\mathscr{L}} = c\frac{\sigma(\lambda)N_0}{\ln 10} \qquad (2.15)$$

and is independent of the absorption pathlength $\mathscr{L}$. For weak absorption, $\mathscr{A} \ll 1$, the right-hand side of Eq. (2.14) can be linearized:

$$\mathscr{T} \approx 1 - \mathscr{A}\ln 10. \qquad (2.16)$$

## 2.3.1 Transmission Spectroscopy

In transmission spectroscopy, the transmitted and incident powers are measured either one after another in two successive scans, or simultaneously, for example with a beam splitter and two detectors. The first approach is susceptible to system drifts that can occur in the time between the two measurements (laser power fluctuations, small misalignments due to thermal expansion, changes in pressure and humidity, wavelength drifts) and should therefore be avoided. But the second case has its problems, too. First, generally only a fraction of $\mathscr{P}_0$ is measured (e.g., 50% if a 50/50 beam splitter is used to split the beam prior to the gas cell). Additionally, the two detectors may not have exactly the same properties, such as responsivity, response time, offset voltage, amplifier gain. Second, absorption due to the gas sample is not the only cause of the decrease of transmitted power. Windows (both of the cell and possibly of the detector) and mirrors (especially important in multipass cells) contribute to loss of laser power. The simplest way of taking these effects into account is by assuming that the measured transmitted $\mathscr{P}_m$ and incident $\mathscr{P}_{m,0}$ powers are given by

$$\begin{cases} \mathscr{P}_m(\lambda) &= \mathscr{P}_0(\lambda)\kappa(\lambda)\mathscr{T}(\lambda) \\ \mathscr{P}_{m,0}(\lambda) &= \mathscr{P}_0(\lambda)\kappa_0(\lambda). \end{cases} \qquad (2.17)$$

Notice that all quantities are possibly time-dependent. $\mathscr{P}_0$ is the laser power, and $\kappa$ and $\kappa_0$ are wavelength and time-dependent factors that take into account detector responsivity, beam splitter ratio, window and mirror losses, etalon fringes, and all possible time-dependent effects such as pointing instabilities, mechanical expansion/contraction due to temperature/pressure change. The division of the two equations in Eq. (2.17) yields the (measurable) ratio $Q$,

$$Q(\lambda) \equiv \frac{\mathscr{P}_m(\lambda)}{\mathscr{P}_{m,0}(\lambda)} = \mathscr{T}(\lambda)\frac{\kappa(\lambda)}{\kappa_0(\lambda)} = \mathscr{T}(\lambda)\mathscr{B}(\lambda), \qquad (2.18)$$

where we introduced the *baseline* $\mathscr{B} \equiv \kappa/\kappa_0$. Determination of the spectrometer's baseline requires a second measurement. If the baseline does not change from measurement to measurement, $\mathscr{B}$ can be measured by filling the cell with a non-absorbing gas ($\mathscr{T} = 1$). The true sample transmittance $\mathscr{T}$ is then the double ratio

$$\mathscr{T}(\lambda) = \frac{Q(\lambda)}{\mathscr{B}(\lambda)}. \tag{2.19}$$

However, if the baseline is not exactly reproducible, then the baseline measurement will yield $\mathscr{B}' \neq \mathscr{B}$: the *measured* baseline $\mathscr{B}'$ is not identical to the unknown baseline $\mathscr{B}$ during the sample measurement. We define the *measured* transmittance $\mathscr{T}'$, as opposed to the true sample transmittance $\mathscr{T}$ [Eq. (2.19)], as

$$\mathscr{T}'(\lambda) \equiv \frac{Q(\lambda)}{\mathscr{B}'(\lambda)}. \tag{2.20}$$

The idea is that the baseline should change as little as possible so that $\mathscr{B} \approx \mathscr{B}'$ and $\mathscr{T}'$ is a good approximation for $\mathscr{T}$. If instead of a sample and a baseline measurement, two baseline measurements $\mathscr{B}, \mathscr{B}'$ are taken, and the first one is considered to be the sample measurement $Q$, then with $\mathscr{T} = 1$ [Eq. (2.19)] and $\mathscr{T}' = \mathscr{B}/\mathscr{B}'$ [Eq. (2.20)] we have

$$\delta\mathscr{T} \equiv \mathscr{T}' - \mathscr{T} = \mathscr{T}' - 1 = \frac{\mathscr{B}}{\mathscr{B}'} - 1. \tag{2.21}$$

Equation (2.21) is the deviation of the measured transmittance $\mathscr{T}'$ from the true sample transmittance $\mathscr{T}$. Notice that all three quantities in Eq. (2.21) are wavelength-dependent. A meaningful quantity in this context is the root mean square value of $\delta\mathscr{T}$:

$$\delta\mathscr{T}_{\mathrm{rms}} \equiv \sqrt{\langle(\delta\mathscr{T})^2\rangle}, \tag{2.22}$$

where the average $\langle\rangle$ is to be taken over all wavelengths. The baseline ratio $\mathscr{B}/\mathscr{B}'$ is a crucial quantity with respect to the accuracy of the spectrometer, while the noise of the measured quantities affects the precision[2]. Whether accuracy or precision limits the sensitivity depends on the width of the absorption features. For narrow and isolated absorption lines (Fig. 2.4, left), the true baseline $\mathscr{B}$ can be extrapolated from the points left and right of the absorption[3], since we know that the true sample transmittance $\mathscr{T}$ there is one (i.e., $Q = \mathscr{B}$, Eq. (2.19)). A second measurement is not needed, and noise (precision of $Q$) limits the sensitivity. For broad spectral features (Fig. 2.4, right), the baseline $\mathscr{B}$ cannot be predicted accurately over the full width

---

[2] A measurement is said to be accurate if the average value of the measured quantity lies close to the true value. A measurement is said to be precise if the variance of the measured quantity is small.

[3] We assume that the baseline is "smooth" over the width of an absorption line.

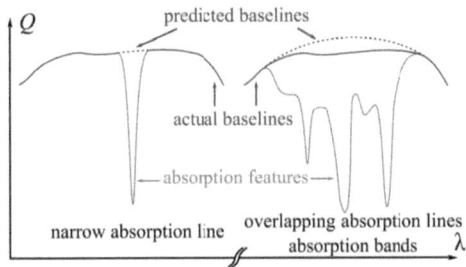

**Figure 2.4** – Narrow (left) and broad (right) absorption features with the same baseline. The hidden part of the baseline was predicted by a cubic polynomial fitted to baseline points left and right of the absorption feature.

of the absorption. A baseline measurement is then necessary and an error $\mathcal{B}/\mathcal{B}' - 1$ is introduced [Eq. (2.21)]. In this case it is the reproducibility (accuracy of $\mathcal{T}'$) that limits the sensitivity.

## 2.3.2 Wavelength Modulation Spectroscopy

Wavelength modulation spectroscopy (WMS) [49, 87–89] is a (nearly) baseline-free technique where the laser wavelength is sinusoidally modulated at a frequency $\mathfrak{f}$ while the average laser wavelength is scanned over an absorption line over a time period which is large compared to $1/\mathfrak{f}$. A detector placed after the gas cell detects the intensity modulation that is produced when the time-varying laser wavelength starts to overlap with the absorption feature. The signal generated by the detector has frequency components at multiples of $\mathfrak{f}$. Without absorption the amplitudes of all harmonics $n\mathfrak{f}$ ($n = 1, 2, \ldots$) are zero, hence there is no baseline. The signal detection at $\mathfrak{f}$ or higher harmonics is also advantageous from the point of view of noise, as the strong low-frequency $1/f$-noise from both laser and detector is filtered out.

The theoretical treatment of WMS can be done both in the time or in the frequency domain. If the modulation frequency is large compared to the half-width of the absorption lines, $\mathfrak{f} \gg \delta v_{\text{line}}$, and the modulation depth is small, we speak of frequency modulation (FM) rather than wavelength modulation. In this case the discussion in the spectral domain is easier, as only a few sidebands of the laser spectrum need to be considered (usually two), only one of which interacts with the absorption line [90]. At low pressure, absorption linewidths are of the order of 200 MHz, so that frequencies in the gigahertz range are needed for FM, which in turn requires very fast detectors and electronics. If the modulation frequency is small compared to the absorption linewidths, $\mathfrak{f} \ll \delta v_{\text{line}}$, we speak of wavelength modulation (WM). WM is usually carried out at frequencies of a few kHz up to tens of kHz.

Schilt and colleagues [91] have given an exhaustive mathematical treatment of WMS for Lorentz-shaped (i.e., pressure broadened) absorption lines that also takes into account the intensity modulation that results in laser diodes and QCLs when the injection current is modulated. We only give a brief overview and a few key results here. The instantaneous laser frequency $v$ is given as

$$v \equiv v_c + \delta v \cdot \sin(2\pi \mathfrak{f} t), \tag{2.23}$$

where $\delta v$ is the modulation depth and $v_c$ is the central laser frequency around which the laser wavelength is sinusoidally modulated. It is convenient to introduce the normalized frequency deviation from the absorption line center:

$$x \equiv \frac{v - v_{\text{line}}}{\delta v_{\text{line}}} \qquad x_c \equiv \frac{v_c - v_{\text{line}}}{\delta v_{\text{line}}}, \tag{2.24}$$

where $v_{\text{line}}$ and $\delta v_{\text{line}}$ are the position and half-width of the absorption line, respectively. With this definition we can rewrite Eq. (2.23) as

$$x = x_c + m \cdot \sin(2\pi \mathfrak{f} t), \tag{2.25}$$

where $m \equiv \delta v / \delta v_{\text{line}}$ is the modulation index. The time-varying incident power $\mathcal{P}(x_c, t)$ when the central laser frequency is held at $x_c$ is equal to the laser power (which we consider constant for now) times the transmittance $\mathcal{T}$:

$$\mathcal{P}(x_c, t) = \mathcal{P}_0 \mathcal{T}(x_c, t) = \mathcal{P}_0 \mathcal{T}(x_c + m \cdot \sin(2\pi \mathfrak{f} t)). \tag{2.26}$$

The transmittance $\mathcal{T}$ can be expanded in a Taylor series:

$$\mathcal{T}(x_c, t) = \mathcal{T}(x_c) + m \mathcal{T}'(x_c) \sin(2\pi \mathfrak{f} t) + \\ + \frac{m^2}{2} \mathcal{T}''(x_c) \sin^2(2\pi \mathfrak{f} t) + \frac{m^3}{6} \mathcal{T}^{(3)}(x_c) \sin^3(2\pi \mathfrak{f} t) + \cdots \tag{2.27}$$

Since $\mathcal{T}(x_c, t)$ is periodic (in $t$) with period $1/\mathfrak{f}$, it can also be written as a Fourier series:

$$\mathcal{T}(x_c, t) = \sum_{k=0}^{\infty} \mathfrak{F}_k\{\mathcal{T}\}(x_c) \cdot \sin(2\pi k \mathfrak{f} t + \phi_k), \tag{2.28}$$

where we have introduced the symbol $\mathfrak{F}_k\{\mathcal{T}\}(x_c)$ for the $k$th Fourier coefficient of $\mathcal{T}$ at the laser frequency $x_c$, and the phase angle $\phi_k$. This representation is a convenient description of $\mathcal{T}$ because the coefficients $\mathfrak{F}_k$ can be measured with lock-in detection. Using basic trigonometric identities the sum in Eq. (2.27) can be rewritten in the form of Eq. (2.28). The first few values of $\mathfrak{F}_k$ and $\phi_k$ are:

$$\mathfrak{F}_0(x_c) = \mathcal{T}(x_c) + \frac{m^2}{4}\mathcal{T}''(x_c) + \frac{m^4}{64}\mathcal{T}^{(4)}(x_c) + \cdots \qquad \phi_0 = -\pi/2$$

$$\mathfrak{F}_1(x_c) = m\mathcal{T}'(x_c) + \frac{m^3}{8}\mathcal{T}^{(3)}(x_c) + \cdots \qquad \phi_1 = 0$$

$$\mathfrak{F}_2(x_c) = -\frac{m^2}{4}\mathcal{T}''(x_c) - \frac{m^4}{48}\mathcal{T}^{(4)}(x_c) + \cdots \qquad \phi_2 = -\pi/2$$

$$\mathfrak{F}_3(x_c) = -\frac{m^3}{24}\mathcal{T}^{(3)}(x_c) + \cdots \qquad \phi_3 = 0 \qquad (2.29)$$

For small modulation indices $m \ll 1$, only the first term needs to be considered for each $\mathfrak{F}_k$: $\mathfrak{F}_0 \sim \mathcal{T}$, $\mathfrak{F}_1 \sim m\mathcal{T}'$, $\mathfrak{F}_2 \sim m^2\mathcal{T}''$, $\mathfrak{F}_3 \sim m^3\mathcal{T}^{(3)}$. This property of WMS has earned it the name *derivative* spectroscopy, since the measured quantities are proportional to the derivatives of the transmittance $\mathcal{T}$. This is no longer true at large modulations $m \gtrsim 1$. The Fourier coefficients $\mathfrak{F}_k$ can be easily computed numerically for a given lineshape function (e.g., Gaussian for Doppler-broadened lines) by taking the Fourier transform of $\mathcal{T}$. The amplitudes of the first ($\mathfrak{F}_1$) and second ($\mathfrak{F}_2$) harmonics for a Doppler lineshape with peak absorbance $10^{-5}$ are given in Fig. 2.5a,b for four different modulation indices $m$. In Fig. 2.5c the peak-to-peak and zero-to-peak amplitudes of the 2f spectra (Fig. 2.5b) are shown as a function of modulation index $m$. The maxima are at $m = 2.1$ and $m = 2.4$ [49, 92]. If laser power modulation induced by injection current modulation should also be taken into account, a more detailed discussion is needed [91]. The upshot is that regardless of the modulation index, the current-induced intensity modulation of the laser distorts the signals at every harmonic. In particular, a signal is produced even without absorption and, consequently, the measurements are no longer baseline-free. This is intuitively clear, since the signals generated by both the absorption and the intensity modulation of the laser appear at the same frequencies $nf$. In practice, 2f detection is usually chosen because $1/f$-noise is smaller compared to 1f detection, the baseline is weaker and the signal is symmetrical with the peak at the position of the absorption line center.

**Comparison of Wavelength Modulation with Direct Transmission**

When the signal produced by the detector is digitized, a *quantization* error is introduced [53]. This error depends on the analog-to-digital converter's (ADC) resolution (number of bits). The advantage of having nearly no baseline is that the full range of the ADC can be used to measure the signal. For

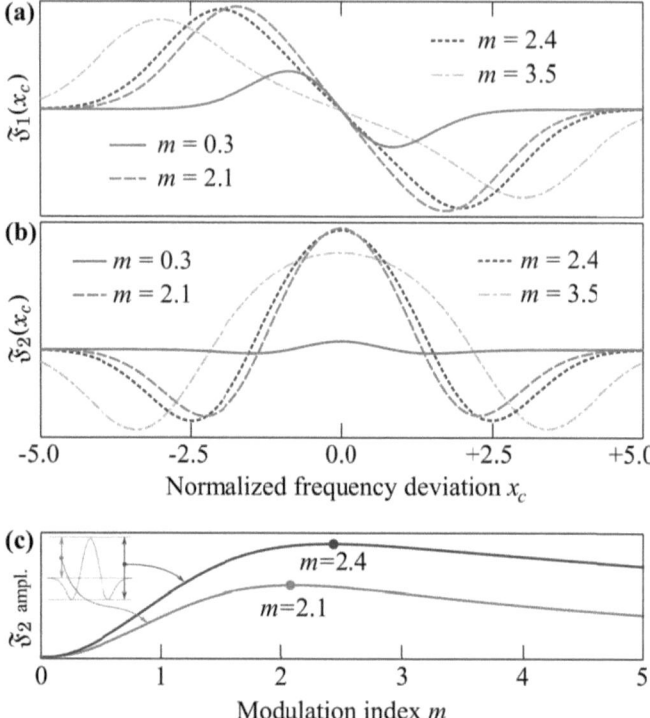

**Figure 2.5** – (a) Amplitude of the first harmonic $\mathfrak{F}_1$ (frequency $\mathfrak{f}$) versus normalized frequency deviation $x_c$ [Eq. (2.24)] for a Gaussian lineshape with peak absorbance $10^{-5}$ for four different modulation indices $m$. (b) Amplitude of the second harmonic $\mathfrak{F}_2$ (frequency $2\mathfrak{f}$) versus normalized frequency deviation $x_c$ for a Gaussian lineshape for four different modulation indices $m$. (c) Maximum amplitude of $\mathfrak{F}_2$ versus modulation index $m$. The two curves indicate the peak-to-peak and zero-to-peak values. The maxima are indicated.

example, a 12 bit ADC has a resolution of $\Delta = 2^{-12} \approx 2.4 \times 10^{-4}$. If a measurement provides a signal with amplitude $10^{-5}$ on top of a baseline with amplitude 1, the ADC cannot distinguish it from the baseline. However, if there is no baseline, the signal could ideally[4] be measured with a relative precision of $2.4 \times 10^{-4}$. This is one of the two arguments in favor of WMS. The other concerns the position of the detection passband: in WMS it is centered around the modulation frequency $\mathfrak{f}$ (or a multiple thereof), whereas in

---

[4]This would probably require some amplification to match the signal amplitude to the input range of the ADC.

direct transmission it is centered around the frequency zero. By averaging several fast sweeps over the absorption line this can be changed and better noise reduction can be achieved (see below).

One could imagine that signals smaller than the ADC resolution cannot be detected under any circumstances, but this is fortunately not true. If the signal is oversampled (i.e., sampled at a rate larger than twice the required measurement bandwidth[5]) and then decimated, the effective number of bits is increased through an effect known as dithering [93]: repeated measurements of the same quantity will yield different outcomes if there is sufficient noise, and the average value of these measurements will be close to the true value of the measured quantity[6]. Thus by decimating or averaging the oversampled signal the quantization error is reduced. The lowest total (quantization plus input) noise is achieved when the noise amplitude is equal to the resolution $\Delta$ of the ADC [94]. If the noise amplitude is smaller than this value, it should be increased artificially. With a 16-bit ADC the resolution is $\Delta = 2^{-16} = 1.5 \times 10^{-5}$: as most real world applications have larger relative noise values it is not often necessary to further increase the noise amplitude, unless when working with very low resolution ADCs. In fact, it may be difficult to achieve noise amplitudes of only $\Delta$.

Lins *et al* [95] have recently shown that despite the strong baseline, direct transmission spectroscopy is only slightly more demanding in terms of ADC resolution compared to WMS, a result that was attributed to the natural dithering of the measured signal due to laser and detector noise. The noise in the signal fed to the ADC should be kept as small as possible (but not smaller than $\Delta$). In WMS this can be achieved by an appropriate choice of lock-in time constant (for a given total measurement time $\mathfrak{T}$) and modulation frequency $\mathfrak{f}$ (since $1/f$-noise is lower at higher frequencies). In direct transmission spectroscopy a different approach is needed. The equivalent of the lock-in's time constant would be a low-pass filter that removes high-frequency components from the measured signal. If the total measurement time is the same for both methods, then the same detection bandwidth $B$ can be chosen for both systems[7]. Thus, while in WMS the passband with width $B$ is centered around the frequency $2\mathfrak{f}$ (for $2\mathfrak{f}$ detection), in direct

---

[5]The Nyquist theorem states that in order to detect frequencies up to $F$ a sampling rate of at least $F_N = 2F$ is required. $F_N$ is the Nyquist frequency.

[6]The ADC can measure values in steps of $\Delta$. If a signal is just slightly below $n\Delta$, the ADC will register the value $n\Delta$ at every sample. However, if noise with amplitude $\Delta$ is present (or otherwise added to the system), the ADC will occasionally register a larger $((n+1)\Delta)$ or smaller $((n-1)\Delta)$ value. How often this happens depends on how close the true value is to $n\Delta$.

[7]Since the $2\mathfrak{f}$ WM signal and the direct signal don't have the same shape, their spectra are different, and the required bandwidth may not be identical. In any case it is of the same order of magnitude.

transmission that same passband is centered around the frequency zero. This obviously results in more noise being transmitted in the latter case. Instead of a single measurement with duration $\mathfrak{T}$, $\mathfrak{N}$ measurements with a shorter duration $\tau \equiv \mathfrak{T}/\mathfrak{N}$ should be carried out and then averaged. "Accelerating" the signals by a factor $\mathfrak{N}$ dilates the original spectrum along the frequency axis by the same factor, and thus increases the required detection bandwidth by a factor $\mathfrak{N}$. Averaging the $\mathfrak{N}$ measured spectra effectively reduces the bandwidth by a factor $1/\mathfrak{N}$, so that the final detection bandwidth is, unsurprisingly, still $B$. But this passband, as shown below, is no longer located (exclusively) around the frequency zero where $1/f$-noise is strong, hence the amount of transmitted noise is smaller. A signal $y(t)$ with bandwidth $B$ measured once over a time $\mathfrak{T}$ can be written as a Fourier integral

$$y(t) = \int_{-B}^{B} \tilde{y}(f) \exp(2\pi i f t) df, \qquad (2.30)$$

with

$$\tilde{y}(f) \equiv \int_{0}^{\mathfrak{T}} y(t) \exp(-2\pi i f t) dt, \qquad (2.31)$$

where the signal $y$ is assumed to be zero outside of $[0, \mathfrak{T}]$ and $\tilde{y}(|f| > B) = 0$. We now make the signal $y$ $\mathfrak{T}$-periodic

$$y_p(t) \equiv \frac{1}{\mathfrak{T}} \sum_{k \in \mathbb{Z}} \hat{y}_k \exp(2\pi i k t/\mathfrak{T}) \qquad (2.32)$$

with

$$\hat{y}_k \equiv \int_{0}^{\mathfrak{T}} y(t) \exp(-2\pi i k t/\mathfrak{T}) dt = \tilde{y}(k/\mathfrak{T}), \qquad (2.33)$$

and then "accelerate" it by a factor $\mathfrak{N}$, so that every measurement lasts only a time $\tau$:

$$z(t) \equiv y_p(\mathfrak{N}t) = \frac{1}{\mathfrak{T}} \sum_{k \in \mathbb{Z}} \hat{y}_k \exp(2\pi i k \mathfrak{N}t/\mathfrak{T}) = \frac{1}{\mathfrak{T}} \sum_{k=-B\mathfrak{T}}^{B\mathfrak{T}} \tilde{y}(k/\mathfrak{T}) \exp(2\pi i k t/\tau). \qquad (2.34)$$

In the last equality in Eq. (2.34) we have used Eq. (2.33) and the fact that $\tilde{y}(f) = 0$ for $|f| > B$. The "faster" signal $z$ is $\tau$-periodic and has bandwidth $B' \equiv B\mathfrak{T}/\tau = B\mathfrak{N}$. Its spectral components are located at multiples of the repetition frequency $f_r \equiv 1/\tau$, and its highest non-zero harmonic is $B\mathfrak{T}$. The averaging of $\mathfrak{N} = \mathfrak{T}f_r$ waveforms over a total time duration $\mathfrak{T}$ corresponds to a filter function [93]

$$|\mathscr{F}(f, \mathfrak{T}, f_r)| \equiv \frac{1}{\mathfrak{T}f_r} \left| \frac{\sin(\pi f \mathfrak{T})}{\sin(\pi f/f_r)} \right|. \qquad (2.35)$$

This function has passbands at $f = \mathbb{Z}f_r$, i.e., at the exact location where the harmonics of the periodic signal $z$ are (Fig. 2.6). There are additional weak side-passbands at $f = \mathbb{Z}/\mathfrak{T}$ but they get smaller if $\mathfrak{N}$ is increased. The width of one passband can be estimated by computing the distance between the nearest two zeroes left and right of a passband and is approximately $1/\mathfrak{T}$. The total bandwidth of this filter is infinite, but since only the first $B\mathfrak{T}$ harmonics must be considered, an additional low-pass filter with a cut-off frequency of $B\mathfrak{T}f_r$ should be used to suppress all passbands that would otherwise just introduce noise. The total detection bandwidth for the signal $z$ after averaging is then $B\mathfrak{T} \cdot (1/\mathfrak{T}) = B$. Although this is the same bandwidth as with a single measurement of duration $\mathfrak{T}$, the difference is that there isn't a passband centered around the frequency zero anymore [Eq. (2.30)]. Instead, several narrower passbands located at $0, f_r, 2f_r, \ldots$ provide a better noise suppression for $1/f$-noise, even though the total bandwidth remains the same. It is clear then, that the passbands should be at frequencies as high as possible where noise is weaker, but this could require unachievable repetition rates. DFB lasers tuned over a single absorption line can be scanned at up to 1–2kHz [95]. In general, WMS provides lower noise levels, especially with lasers or detectors very noisy at low frequencies. The quan-

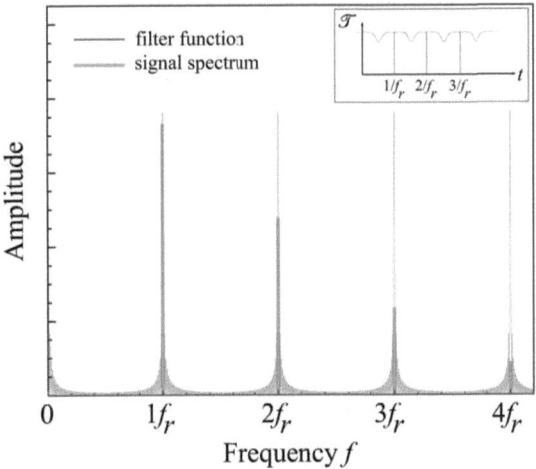

**Figure 2.6** – Filter function equivalent to the averaging of 10 periods and amplitudes of the first four harmonics of the transmittance function $\mathcal{T}$ for a Doppler-broadened absorption line if the measurement is repeated at the repetition frequency $f_r$ (see inset). The extremely strong zero-frequency component (which provides the baseline value 1 of the transmittance) has been omitted.

tity relevant for the sensitivity, however, is usually the signal-to-noise ratio (SNR), and since the direct signal is stronger compared to the 2f amplitude $\mathfrak{F}_2$ of WMS by a factor of 1/0.438 [96], it is not clear *a priori* which detection scheme provides best sensitivity. This question can only be answered by careful examination of the noise profiles (laser and detector) including quantization noise [95].

### 2.3.3 Signal-to-noise Ratio with Multipass Cells

For both wavelength modulation and direct transmission spectroscopy, when the laser is scanned over an absorption line a signal is generated. In direct transmission this signal, $\mathscr{S}_d$, is caused by a change in incident power on the detector due to the absorption:

$$\mathscr{S}_d \equiv \mathscr{P}_0(1-\mathscr{T}_0), \tag{2.36}$$

where $\mathscr{P}_0$ is the laser power and $\mathscr{T}_0 \equiv \mathscr{T}(x_c = 0)$ [Eq. (2.24)] is the transmittance at the center of the absorption line. In wavelength modulation spectroscopy the transmitted power $\mathscr{P}_0 \mathscr{T}(x_c, t)$ is time-dependent due to the laser frequency being oscillated at a frequency f [Eq. (2.23)]. The relevant signal $\mathscr{S}_w$ is the amplitude of the second harmonic of $\mathscr{P}_0 \mathscr{T}$, at its peak $x_c = 0$ (Fig. 2.5b):

$$\mathscr{S}_w \equiv \mathfrak{F}_2\{\mathscr{P}_0 \mathscr{T}_0\} = \mathscr{P}_0 \mathfrak{F}_2\{\mathscr{T}_0\}. \tag{2.37}$$

In Fig. 2.7 the power-normalized signal amplitudes $\mathscr{S}_d/\mathscr{P}_0$ and $\mathscr{S}_w/\mathscr{P}_0$ [Eqs. (2.36), (2.37)] are shown as a function of absorbance $\mathscr{A} = -\log_{10} \mathscr{T}$ [Eq. (2.13)] for a Gaussian-shaped absorption line (modulation index $m = 2.1$). As was previously known [49], the 2f amplitude $\mathscr{S}_w$ is smaller by a factor 0.438 compared to the signal achieved with direct transmission, $\mathscr{S}_d$. Therefore, if both systems have the same amount of noise – this is possible if the noise is mainly "white" (see Sec. 2.3.2), i.e., with no frequency dependence – direct transmission will provide a better SNR [95]. It also follows that $\mathscr{S}_d$ and $\mathscr{S}_w$ are proportional to each other and, with Eqs. (2.36) and (2.37),

$$\mathfrak{F}_2\{\mathscr{T}_0(t)\} \propto 1 - \mathscr{T}_0. \tag{2.38}$$

The slope of both curves $\mathscr{S}_d$ and $\mathscr{S}_w$ is one, meaning that both $\mathscr{S}_d$ and $\mathscr{S}_w$ are proportional to the absorbance $\mathscr{A}$ over at least nine orders of magnitude, from $10^{-10}$ to $10^{-1}$ absorbance units. If the laser power $\mathscr{P}_0$ can be measured, then both detection schemes allow direct computation of the transmittance:

$$\mathscr{T}_0 = 1 - \frac{\mathscr{S}_d}{\mathscr{P}_0} \tag{2.39}$$

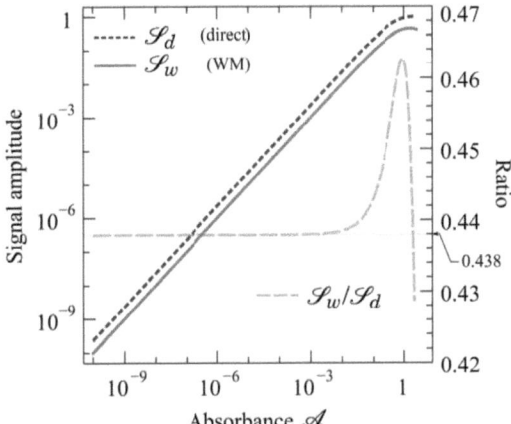

**Figure 2.7** – Computed signal amplitudes for direct ($\mathscr{S}_d$) and wavelength modulation (modulation index $m = 2.1$) spectroscopy ($\mathscr{S}_w$) versus absorbance (for a Gaussian-shaped absorption line), and ratio $\mathscr{S}_w/\mathscr{S}_d$.

and
$$\mathscr{T}_0 = 1 - \frac{\mathscr{S}_w}{0.438 \cdot \mathscr{P}_0}. \tag{2.40}$$

Notice that the additional factor in Eq. (2.40) is strongly dependent on the modulation index and the value 0.438 is valid only for $m = 2.1$, for which the zero-to-peak height of the 2f amplitude of a Gaussian lineshape is maximal (Fig. 2.5). In reality, it is more accurate to measure its value with a calibration. At high absorbance $\mathscr{A}$ the linearity is not given anymore for neither $\mathscr{S}_d$ nor $\mathscr{S}_w$.

High absorbances, while still remaining in the linear range (Fig. 2.7), are desirable since they provide larger and therefore more easily measurable signals. A common way to achieve that is by employing multipass cells [54] in order to increase the absorption pathlength $\mathscr{L}$ [Eq. (2.13)]. If $\mathscr{T}_1$ denotes the transmittance at the peak of the absorption line for a single pass through a multipass cell, $n$ is the total number of passes, and $\mathscr{R}$ is the mirror reflectivity (assumed to be the same for both mirrors), then Eqs. (2.36) and (2.37) can be rewritten as

$$\mathscr{S}_d(n) = \mathscr{P}(1 - \mathscr{T}_1^n) = \mathscr{P}_0 \mathscr{R}^{n-1}(1 - \mathscr{T}_1^n) \tag{2.41}$$

and
$$\mathscr{S}_w(n) = \mathscr{P}\mathfrak{F}_2\{\mathscr{T}_1^n\} = \mathscr{P}_0 \mathscr{R}^{n-1} \mathfrak{F}_2\{\mathscr{T}_1^n\}, \tag{2.42}$$

where $\mathscr{P}_0$ is the laser power entering into the multipass cell and $\mathscr{P}_0 \mathscr{R}^{n-1}$ is the power transmitted through the cell if the laser wavelength is not on the absorption line. The signal-to-noise ratio (SNR) is defined as

$$\text{SNR}_d(n) \equiv \frac{\mathscr{S}_d(n)}{\mathscr{N}(n)} \qquad \text{SNR}_w(n) \equiv \frac{\mathscr{S}_w(n)}{\mathscr{N}(n)}, \qquad (2.43)$$

where $\mathscr{N}$ is the noise. Noise can be considered to consist of two parts: one part (like laser noise) that is proportional to the power incident on the detector,

$$\mathscr{N}_p = \kappa \mathscr{P}_0 \mathscr{R}^{n-1} \qquad (2.44)$$

and a constant part from the detector and electronics

$$\mathscr{N}_c = \text{const.} \qquad (2.45)$$

The overall SNR can then be written in the form

$$\text{SNR}_d(n) = \frac{\mathscr{S}_d(n)}{\sqrt{\mathscr{N}_p^2 + \mathscr{N}_c^2}} \propto \frac{1 - \mathscr{T}_1^n}{\sqrt{\kappa^2 + (\beta \mathscr{R}^{1-n})^2}} \qquad (2.46)$$

and

$$\text{SNR}_w(n) = \frac{\mathscr{S}_w(n)}{\sqrt{\mathscr{N}_p^2 + \mathscr{N}_c^2}} \propto \frac{\mathfrak{F}_2\{\mathscr{T}_1^n\}}{\sqrt{\kappa^2 + (\beta \mathscr{R}^{1-n})^2}} \qquad (2.47)$$

with $\beta \equiv \mathscr{N}_c/\mathscr{P}_0$ the detector noise relative to the laser power measured at the input of the multipass cell. Two special cases of Eqs. (2.46) and (2.47) can be distinguished: only proportional noise ($\beta = 0$) and only constant noise ($\kappa = 0$). In the first case, the SNR is given by the numerators of Eqs. (2.46) and (2.47), respectively (since $\kappa$ is a constant). The mirror reflectivity $\mathscr{R}$ has no effect, and more passes $n$ lead to a better SNR:

$$\text{SNR}_d^{(p)}(n) \propto 1 - \mathscr{T}_1^n \qquad (2.48)$$

and

$$\text{SNR}_w^{(p)}(n) \propto \mathfrak{F}_2\{\mathscr{T}_1^n\} \propto 1 - \mathscr{T}_1^n. \qquad (2.49)$$

The last proportionality in Eq. (2.49) follows from Eq. (2.38). In the second case, the SNRs are given by

$$\text{SNR}_d^{(c)}(n) \propto \mathscr{R}^{n-1}(1 - \mathscr{T}_1^n) \qquad (2.50)$$

and

$$\text{SNR}_w^{(c)}(n) \propto \mathscr{R}^{n-1} \mathfrak{F}_2\{\mathscr{T}_1^n\} \propto \mathscr{R}^{n-1}(1 - \mathscr{T}_1^n). \qquad (2.51)$$

The second factor in Eqs. (2.50) and (2.51) is proportional to $n$ (for weak absorption), so that $\text{SNR}^{(c)} \sim n\mathscr{R}^{n-1} \to 0$ ($n \to \infty$) since $\mathscr{R} < 1$. There is a number of passes $n^*$ which provides the highest SNR: it can be estimated by differentiating Eqs. (2.50) and (2.51) with the assumption that the absorption is weak $(1 - \mathcal{T}_1^n \approx n(1 - \mathcal{T}_1))$ and setting the derivative equal to zero:

$$n^* = \frac{-1}{\ln \mathcal{R}}. \qquad (2.52)$$

This holds true for the case $\kappa = 0$. In general, both kind of noise (proportional and constant) are present. Then, $n^*$ is shifted towards higher values. The above is clarified in Fig. 2.8. With exclusively proportional (laser) noise, the effect of the finite reflection losses of the mirrors is unimportant because the noise as well as the power are both attenuated in the same manner by the passage through the multipass cell. The SNR increases with $n$ regardless of the mirror reflectivity. With exclusively constant (detector) noise, the SNR increases at first because the loss of power, $\mathcal{R}^{n-1}$, is small compared to the gain in $(1 - \mathcal{T}_1^n) \propto n$ [Eqs. (2.50) and (2.51)]. However, as the losses increase exponentially, the signal will eventually decrease. The number of passes that yields best SNR is given in Eq. (2.52). If both proportional and constant noise are present, the situation is somewhere in between the two extreme cases above, with the SNR increasing up to an optimal number of passes and then decreasing quickly. The percentages in Fig. 2.8 are the mirror reflectivities $\mathcal{R}$. For worse mirrors, the optimum number of passes is reached at lower values of $n$, because the power attenuation per passage is larger.

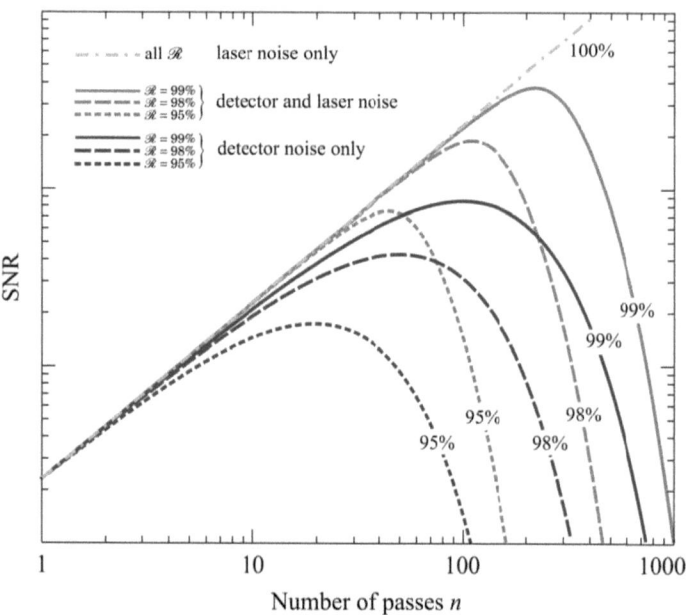

**Figure 2.8** – Signal-to-noise ratio (SNR, logarithmic scale) for three different noise scenarios as a function of the number of passes $n$ through a multipass cell. The mirrors are assumed to have reflectivities of $\mathscr{R} = 100\%, 99\%, 98\%$ and $95\%$, respectively. Notice that for laser noise only, all curves coincide, independently of $\mathscr{R}$. Notice also, that for $\mathscr{R} = 100\%$ all curves coincide with the laser noise only curve, independently of the type of noise.

## 2.4 Difference Frequency Generation Based Laser Spectrometer

### 2.4.1 Setup

A schematic representation of the DFG spectrometer is given in Fig. 2.9. The signal laser is a fiber-coupled external cavity diode laser (ECDL, Santec Corp., Japan, model TSL-210) tunable from 1520 to 1600 nm with a linewidth < 1 MHz and an output power of at least 5 mW. About 1% of the power is coupled into a wavemeter (Burleigh, U.S.A., model WA-1100, via a polarization-maintaining (PM) filter coupler (Opto-Link Corp. Ltd., Hong Kong) for wavelength monitoring. A $f = 400$ mm lens is placed about 3 cm after the fiber output coupler, followed by a half-wave plate used to rotate the polarization in the vertical direction. The pump laser is a passively Q-switched Nd:YAG laser (InnoLight GmbH, Germany, model M800,) with a repetition rate of 4–8 kHz and 6 ns pulses. It operates at 1064.5 nm with an average output power around 300 mW and pulse peak powers of the order of 5 kW. A collimating $f = 200$ mm lens follows a quarter-wave and half-wave plate, needed to linearize and rotate, respectively, the plane of polarization of the Nd:YAG laser beam. Pump and signal beam are combined with a dichroic mirror prior to a $f = 75$ mm lens. A small fraction of the pump beam traverses the dichroic mirror and is captured by a fast Si photodiode (Hamamatsu, Japan, model S4753) which provides the trigger pulse for the data acquisition. A 50 mm long, 10 mm wide, 0.5 mm thick undoped PPLN crystal (Crystal Technology, U.S.A.) with 8 different poling periods (from 28.5 to 29.9 $\mu$m in steps of 0.2 $\mu$m) is held inside an oven (Super Optronics, U.S.A.). The pump and signal laser wavelength determine which poling period should be used and what temperature should be set (see Sec. (2.4)). The polarization is vertical for all three beams (pump, signal and idler). A $f = 99$ mm $CaF_2$ lens collimates the idler after the PPLN crystal. A germanium filter removes the pump and signal beam. The idler is coupled into the multipass cell with a $f = 379$ mm and $f = 300$ mm $CaF_2$ lens. The Herriott-type home-made high temperature multipass cell (HTMC) [24, 25] is a stainless steel cylinder with 2 $\ell$ volume provided with bellows for compensation of thermal expansion, and internal metallic mirrors. The longest absorption pathlength is 35 m and can be adjusted to lower values by changing the mirror separation distance. The HTMC can be evacuated with a rotary (Alcatel, France) and a turbo pump (Leybold-Heraeus, Germany, Turbovac 50) down to a pressure of $10^{-2}$–$10^{-3}$ mbar. About 10% of the power is reflected by a beam splitter and focused by a $f = 200$ mm $CaF_2$ lens onto the reference detector. The beam transmitted

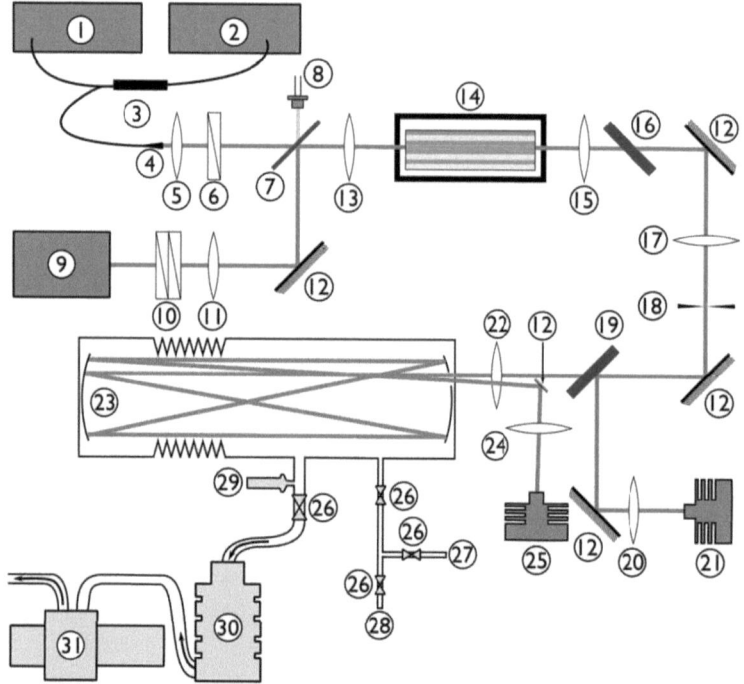

**Figure 2.9** – Schematic drawing of the DFG spectrometer. (1) CW external cavity diode laser (ECDL, 1520–1600 nm, 5 mW power); (2) wavemeter for ECDL; (3) polarization-maintaining (PM) fiber with filter coupler; (4) fiber output coupler; (5) $f = 400$ mm lens; (6) half-wave plate; (7) dichroic mirror; (8) Si photodiode; (9) Q-switched Nd:YAG laser (1064.5 nm, 8 kHz repetition rate; 5 kW peak power; 300 mW average power); (10) half and quarter-wave plate; (11) $f = 200$ mm lens; (12) mirrors; (13) $f = 75$ mm lens; (14) crystal oven with periodically poled lithium niobate (PPLN) crystal (8 poling periods 28.5–29.9 µm in steps of 0.2 µm, 50 mm long, 10 mm wide, 0.5 mm thick); (15) $f = 99$ mm $CaF_2$ lens; (16) germanium filter; (17) $f = 379$ mm $CaF_2$ lens; (18) variable diameter aperture; (19) $CaF_2$ beam splitter; (20) $f = 200$ mm $CaF_2$ lens; (21) reference detector; (22) $f = 300$ mm $CaF_2$ lens; (23) high-temperature multipass cell (HTMC, max. path length 35 m); (24) $f = 100$ mm $CaF_2$ lens; (25) transmission detector; (26) valves; (27) connection to gas bottles (nitrogen, carbon dioxide); (28) connection to sample bags; (29) pressure gauge; (30) turbopump; (31) rotary pump.

through the HTMC is focused by a $f = 100$ mm $CaF_2$ lens onto the transmission detector. The two detectors are two-stage thermo-electrically cooled and preamplified photo-voltaic (HgCdZn)Te detectors optimized for operation at 4 $\mu$m with a peak responsivity of 120 kV/W of incident power (Vigo Systems SA, Poland, model PDI-2TE-4/VPDC-0.1i).

The key features of the spectrometer are described later in Sec. 2.4.4 and in Tab. 2.1.

### 2.4.2 Tuning

Tuning of the idler wavenumber is achieved by tuning the wavelength of the ECDL (signal laser) while adapting the crystal temperature $T$ so that the QPM condition Eq. (2.7) remains fulfilled. If the temperature-dependence of the refractive index and the thermal expansion properties of the crystal are known, the temperature for which QPM occurs can be computed with Eq. (2.7). Another way is to measure the idler power as a function of PPLN temperature for some predefined signal wavelengths (Fig. 2.10a). The signal wavelength for which the maximum idler power is achieved for a given PPLN temperature can then be computed (Fig. 2.10b). A cubic polynomial,

$$\lambda_s^{\text{opt}}(T) = a_0 + a_1 T + a_2 T^2 + a_3 T^3, \qquad (2.53)$$

can be used to fit the data points and provides a formula to compute the optimum signal wavelength for any PPLN temperature.

When measuring a spectrum, the signal laser wavelength (and the PPLN crystal temperature) can be tuned from low to high or high to low values. Figure 2.11a shows the idler power as a function of signal laser wavelength when tuning from low to high and high to low wavelengths, respectively. The HTMC was filled with argon at 200 mbar. Strong interference fringes appear throughout the tuning range. They make it impossible to fit the baseline with a simple polynomial. Moreover, since the fringe pattern is not exactly reproducible, each sample measurement must be followed (or preceded) by a reference measurement (see Sec. (2.3.1)). There is a significant hysteresis below 1580 nm which heavily affects the detector signal ratio $Q$ (Fig. 2.11b). Possible explanations for the hysteresis could be thermal expansion of the oven and crystal holder, or photorefractive damage in the PPLN crystal [97]. The spectrometer should therefore always be tuned in the same direction (e.g., from low to high signal laser wavelengths), as the baselines for the two directions are totally different. If we repeat the scan from low to high signal wavelength and stop it at $\lambda_s = 1570$ nm by keeping the signal wavelength and the PPLN temperature constant, then the idler power does not remain constant, as is expected, but decreases exponentially with a decay time of $\tau = 220$ s (Fig. 2.12). The final value of the

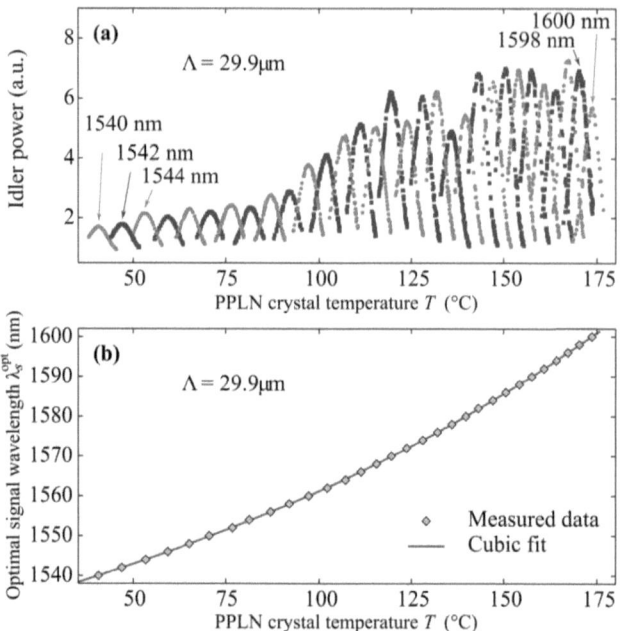

**Figure 2.10** – (a) Idler power measured as a function of PPLN temperature for fixed signal wavelengths from 1540 to 1600 nm in steps of 2 nm. (b) Signal wavelength for which the idler power is maximal as a function of PPLN temperature.

idler power is approximately equal to the power obtained at $\lambda_s = 1570$ nm when scanning from high to low signal laser wavelengths (blue circles in Fig. 2.11). Although the idler power decreases dramatically, $Q$ [Eq. (2.18)] increases only slightly (about 1%). We believe this to be caused by the ongoing heating of the oven insulation and holder – even though the crystal itself is at constant temperature – which could cause slight variations in the PPLN position and orientation. Photorefractive damage could also play a role by distorting the beam profile [98]. The effect shown in Fig. 2.12 can be thought of as a time-dependency in the baseline $\mathscr{B}$ [Eq. (2.18)].

The HTMC was built for operation at elevated temperatures [24]. The bellows that compensate for thermal expansion and keep the mirror separation distance constant while heating make the cell susceptible to changes in atmospheric or internal pressure. This is illustrated in Fig. 2.13, where the relative change, $\delta Q/Q_0$, in detector signal ratio $Q$ with respect to its value $Q_0$ at the initial pressure of 945 mbar is shown. The pressure was decreased

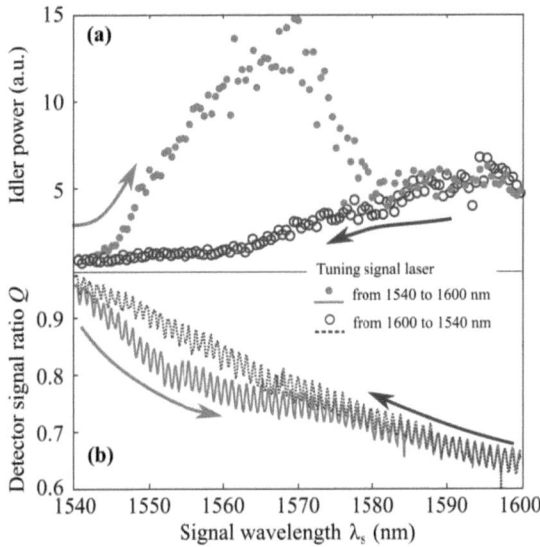

**Figure 2.11** – **(a)** Idler power while tuning the signal laser from low to high (red dots) or high to low (blue circles) wavelengths, with the HTMC filled with argon at 200 mbar. **(b)** Detector signal ratio for the low to high (solid red) and high to low (dashed blue) tuning.

six times almost instantaneously in steps of 5 mbar by quickly opening a valve to which the rotary pump was connected. The experimental data are approximated by the straight line which corresponds to a decrease in $\delta Q/Q_0$ of 0.3% per mbar. We can assume that if the pressure within the cell is kept constant and the atmospheric pressure changes, something similar would be observed. A realistic increase or decrease in atmospheric pressure of about 5 mbar during the sample and reference measurement would then introduce an error of 1.5%. Thus, it is important that both sample and reference measurement are carried out at the same (internal) pressure, and that the time delay between the two measurements is as small as possible to minimize changes in atmospheric pressure.

We conclude that for best reproducibility of $\mathscr{B}$ all measurements should be performed: (a) in the same direction (e.g., from low to high signal laser wavelength); (b) following a predefined *timetable* so that effects that depend on the duration (speed) of the scan – such as the one shown in Fig. 2.12 – can be minimized; (c) the time delay between the sample and reference measurement should be as small as possible, and the pressure and temperature during both measurements should be exactly the same.

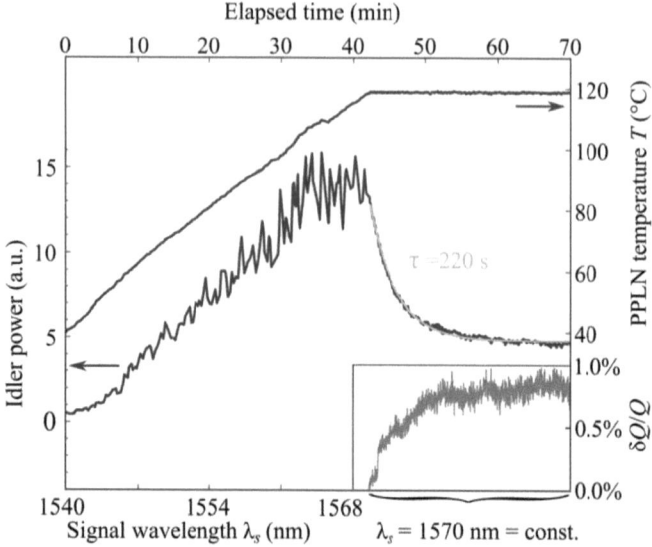

**Figure 2.12** – Idler power during a scan from low to high signal laser wavelengths. The scan is stopped at $t = 42$ min and the PPLN temperature kept constant at 119 °C. Inset: relative deviation of $Q$ from its value at $t = 42$ min.

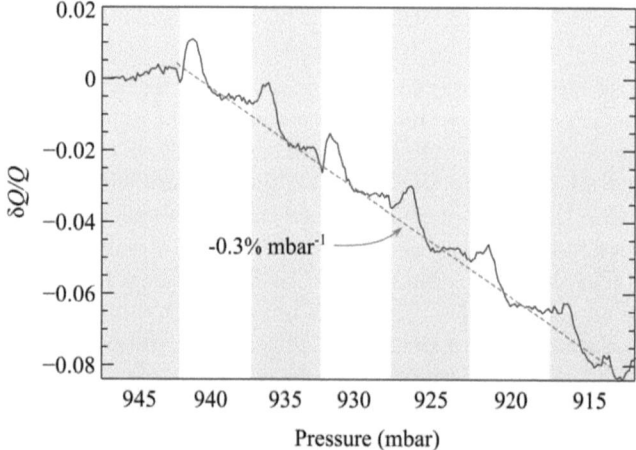

**Figure 2.13** – Relative change in detector signal ratio while step-wise reducing the pressure inside the HTMC. The experimental data (solid) is approximated by a straight line (dashed) with a slope of $-0.3\%$ mbar$^{-1}$.

To ensure that all spectra are recorded under the same conditions – thereby minimizing changes in the baseline $\mathscr{B}$ [Eq. (2.18)] from measurement to measurement – we perform the scan by starting at a fixed PPLN crystal temperature and then increasing it at a rate $r$, so that the crystal experiences the same temperature-time curve at every measurement:

$$T(t) = T_0 + r \cdot t, \qquad (2.54)$$

where $t$ is the elapsed time and $T_0$ is the temperature at $t = 0$. To ensure that $dT/dt = r = $ const. throughout the measurement, the heating of the crystal should start a few degrees below $T_0$. Once $T$ crosses $T_0$ the timer $t$ is started and the measurement begins. The rate $r$ determines the duration of the scan. Larger rates mean less time for data acquisition and averaging and, therefore, more noise. Furthermore, the heating power of the oven is limited. Small rates lead to unacceptably long measurements. A good value for $r$ is 2 °C/min, which results in measurements of 67 min duration (full scan). The reproducibility of the PPLN crystal temperature-time profile is good: the maximum deviation from measurement to measurement is 0.3 °C and the root mean square (r.m.s.) value of the deviation is 0.085 °C. Knowing how much time elapsed since the measurement was started, we can predict (with 0.085 °C margin of error) the temperature of the PPLN crystal at that time and, with Eq. (2.53), the signal laser wavelength that should be set at that time:

$$\lambda_s(t) \equiv \lambda_s^{\text{opt}}(T(t)) = \lambda_s^{\text{opt}}(T_0 + r \cdot t). \qquad (2.55)$$

This can be done in real time or *a priori*; in the latter case Eq. (2.55) is evaluated for some predefined times $t$ (e.g., every 500 ms) and the result is inserted into a table. We call this tuning procedure *timetable* tuning [99], since PPLN crystal temperature, signal laser wavelength and (consequently) idler wavelength follow a predefined timetable.

The recording of a spectrum is schematized in Fig. 2.14. The PPLN crystal temperature $T$ follows a linear profile with slope $r$ [Eq. (2.54)] throughout the measurement (Fig. 2.14a). The signal wavelength $\lambda_s$ is set according to Eq. (2.55) (Fig. 2.14b) and is kept constant during the acquisition (Fig. 2.14c). Once the wavelength is stable and has been measured with the wavemeter, the signals generated on the two detectors by a single laser pulse are acquired, integrated, and the ratio $Q$ [Eq. (2.64)] is computed. These steps are repeated 500 times and the 500 values of $Q$ are averaged and stored. The signal wavelength is then set to the next value and the acquisition cycle begins anew. The choice of acquiring $\mathfrak{N} = 500$ pulses is arbitrary. For large $\mathfrak{N}$ the noise in $Q$ is lower but the acquisition takes longer, which results in fewer points in the spectrum. The slope $r$ can be

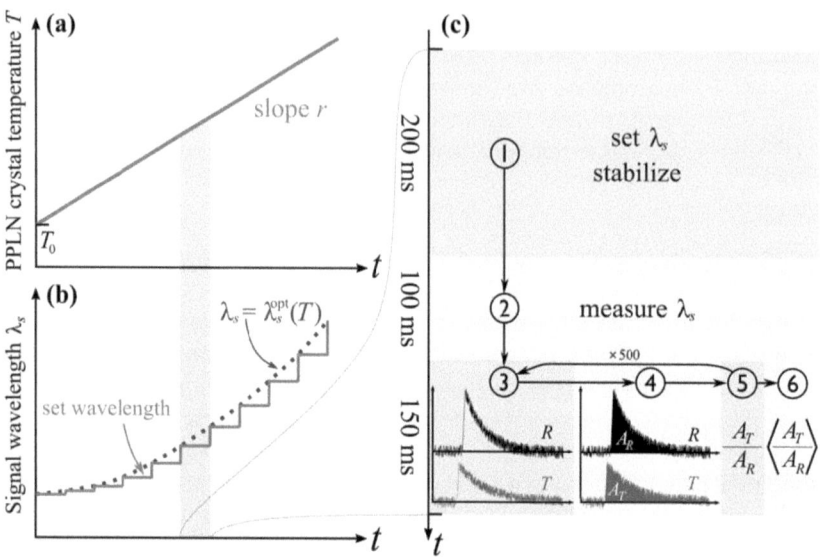

**Figure 2.14** – Recording of a spectrum with the DFG spectrometer. (a) The PPLN crystal temperature $T$ follows a linear profile. (b) The signal wavelength $\lambda_s$ is set according to the current value of $T$ [Eq. (2.55)] and is kept constant during each acquisition cycle. (c) Acquisition cycle: (1) set $\lambda_s$ to the correct value and wait for the laser to stabilize (200 ms); (2) measure the signal wavelength (100 ms); (3) acquire the signal generated by one laser pulse from the reference and transmission detectors; (4) compute the areas $A_R$ and $A_T$ [Eq. (2.57)]; (5) compute the ratio $Q$ [Eq. (2.64)]; the steps (3)–(5) are repeated 500 times; (6) average the $Q$ values. The entire acquisition cycle lasts 450 ms.

reduced to compensate, but this then results in a longer measurement time. With $\mathfrak{N} = 500$ the acquisition of the pulses including processing (integrating, dividing, averaging and storing) lasts about 150 ms (repetition rate of the laser: ∼ 5 kHz). Setting the signal wavelength and waiting until the signal laser power and wavelength are stable takes 200 ms; measuring the wavelength with the wavemeter takes about 100 ms. The entire acquisition cycle lasts 450 ms. The idler can be tuned over 244 cm$^{-1}$ without changing the poling period of the PPLN crystal by tuning the PPLN crystal temperature $T$ from 40 to 173 °C, which at 2 °C/min takes 66.5 min. At 450 ms per point this results in a spectrum with 8870 spectral points (for 244 cm$^{-1}$). The average wavenumber step size is then 0.028 cm$^{-1}$ (830 MHz).

## 2.4.3 Data Acquisition and Evaluation

The voltage pulses generated by the two detectors are digitized by an analog-to-digital converter (ADC, Gage, U.S.A., model CS14100). The acquisition is triggered by the signal provided by the Si photodiode. Sampling occurs at a rate of 50 MHz ($\Delta t = 20\,\text{ns}$) and a resolution of 14 bit. The detectors and the built-in amplifiers have a bandwidth of only 100 kHz, so that the nanosecond laser pulses produce a fast rise of the detector signal (within 1 sample point, 20 ns), followed by an exponential decay with a time constant of $\tau \approx 6\,\mu\text{s}$ (about 300 sample points at 50 MHz sampling rate). The total pathlength $\mathscr{L}$ of the beam inside the HTMC can be determined by measuring the time delay between the arrival of the two pulses on their respective detectors.

It is convenient, in terms of signal-to-noise ratio (SNR), to integrate the pulse shapes instead of only recording the peak values. Consider the sampled signal $\boldsymbol{s} = (s_1, s_2, \ldots)$ with peak amplitude 1 and noise $\sigma$. The SNR of the peak is given by

$$\text{SNR}_p = \frac{1}{\sigma}, \tag{2.56}$$

where the subscript p stands for peak. If instead we integrate the signal $\boldsymbol{s}$ by computing the sum

$$A(n) \equiv \sum_{k=1}^{n} s_k \tag{2.57}$$

with noise $\sigma_A(n) = \sigma\sqrt{n}$, then the SNR becomes

$$\text{SNR}_A(n) = \frac{A(n)}{\sigma\sqrt{n}}. \tag{2.58}$$

Of interest is the SNR with integration relative to the SNR of the peak:

$$R \equiv \frac{\text{SNR}_A(n)}{\text{SNR}_p} = \frac{A(n)}{\sqrt{n}}. \tag{2.59}$$

In Fig. 2.15 the ratio $R$ is plotted for a rectangular-shaped and an exponentially decaying signal. For the rectangular-shaped signal with a fixed pulse width $w$, $R$ is given by

$$R = \kappa(w) \cdot \begin{cases} \sqrt{n/w}, & n/w \leq 1 \\ \sqrt{w/n}, & n/w > 1 \end{cases} \tag{2.60}$$

The factor $\kappa$ depends on $w$, but not on $n$. $R$ is maximal at $n/w = 1$, since integrating over more points further increases the noise but not the signal

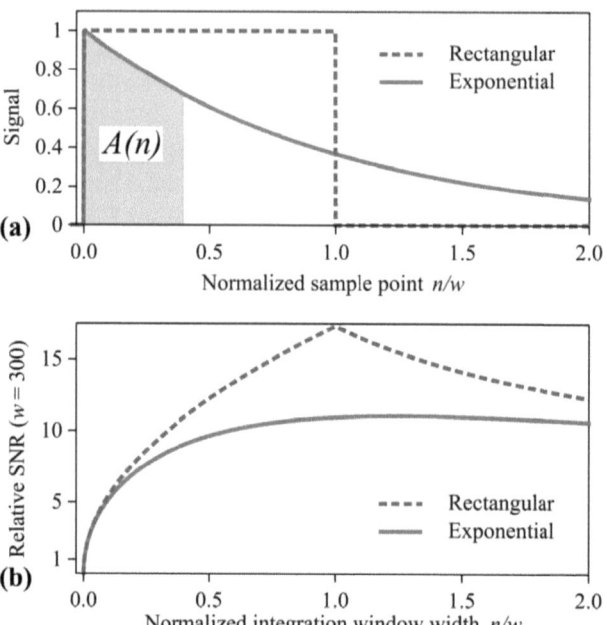

**Figure 2.15** – **(a)** Rectangular-shaped and exponentially decaying signal. **(b)** Relative signal-to-noise ratio [Eq. (2.59)] for a pulse width $w$ of 300 sample points.

area. For the exponentially decaying signal with a fixed decay time $w$, as in our case, $R$ is given by

$$R = \kappa(w) \frac{1 - \exp(-n/w)}{\sqrt{n/w}} \tag{2.61}$$

with

$$\kappa(w) \equiv \frac{1}{\sqrt{w}[1 - \exp(-1/w)]}. \tag{2.62}$$

The maximum of $R$ is at at $n/w \approx 1.26$. For pulses with decay times longer than 100 sample points, the exponential function in Eq. (2.62) can be linearized ($\exp(-1/w) \approx 1 - 1/w$) and $\kappa$ simplifies to $\kappa(w) \approx \sqrt{w}$. The peak value of $R$ can then be written as

$$R^{\max} \approx 0.638\sqrt{w}. \tag{2.63}$$

Choosing a larger integration window does not heavily affect $R$, as can be seen from Fig. 2.15b.

In analogy to Eq. (2.18), we define the detector signal ratio $Q$ as

$$Q(\lambda) \equiv \frac{A_T(\lambda)}{A_R(\lambda)}, \qquad (2.64)$$

where $A_T$ and $A_R$ are the areas [Eq. (2.57)] of the reference (R) and transmission (T) signals, respectively. The SNR of $Q$ is

$$\text{SNR}_Q = \frac{\text{SNR}_A}{\sqrt{2}} = R\frac{\text{SNR}_p}{\sqrt{2}}, \qquad (2.65)$$

where the definition of $R$ [Eq. (2.59)] has been used and it was assumed that both $A_T$ and $A_R$ have the same SNR and are uncorrelated. Measuring $Q$ $\mathfrak{N}$ times and averaging further increases the SNR by a factor $\sqrt{\mathfrak{N}}$:

$$\text{SNR}_{\langle Q \rangle} = R\sqrt{\frac{\mathfrak{N}}{2}}\text{SNR}_p. \qquad (2.66)$$

For a fixed $\mathfrak{N}$, the largest SNR is achieved when the maximum value for $R$ is chosen. From this point on it is understood that $Q$ is always an average value and we will omit the average brackets $\langle \rangle$ for simplicity.

As the two areas $A_T$ and $A_R$ are both proportional to the pulse energy incident on the respective detector, $Q$ is proportional to the sample transmittance $\mathcal{T}$. Ideally, $Q$ would only depend on the wavelength, but the data in Fig. 2.16 shows that this is not the case. When keeping the wavelength constant and tuning the PPLN temperature away from its optimum, a decrease in idler power is expected and observed because the QPM condition [Eq. (2.7)] is no longer fulfilled. The detector signal ratio $Q$ changes as well and this has important consequences regarding the reproducibility of measurements. If in two consecutive scans (of the same sample) the value of $Q$ is measured at the same wavelength but at slightly different PPLN temperatures (or at slightly different laser powers[8]), the results will be different. This effect can be attenuated by the insertion of an aperture along the path of the idler (Fig. 2.9, (18)). We believe that the PPLN temperature affects the spatial beam profile of the idler and, by consequence, the amount of light that reaches the detectors. All measurements made with the DFG spectrometer were carried out with the aperture in place.

A non-absorbing sample (nitrogen) was measured twice consecutively (Fig. 2.17). Since the transmittance $\mathcal{T}$ of the sample is 1 everywhere, the detector signal ratio $Q$ in Fig. 2.17 is the baseline $\mathcal{B}$ of the spectrometer [Eq. (2.18)]. In addition to a negative slope, a set of two fringes can be

---

[8]We have also observed that $Q$ depends on the pump and signal laser power.

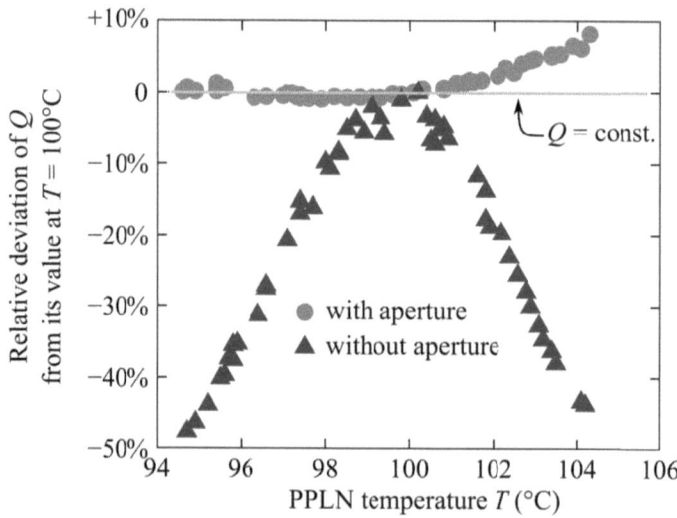

**Figure 2.16** – Relative deviation of the detector signal ratio $Q$ versus PPLN temperature with and without aperture ((18) in Fig. 2.9).

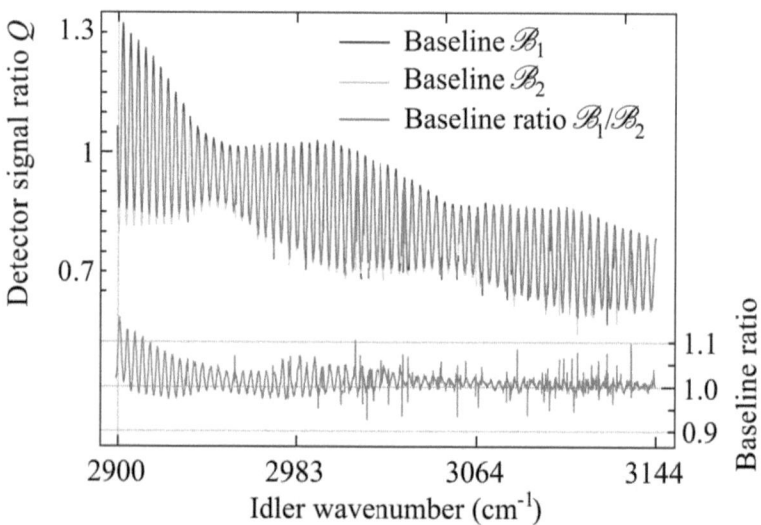

**Figure 2.17** – Two consecutively measured baselines $\mathcal{B}_1, \mathcal{B}_2$ (cell filled with nitrogen) and baseline ratio $\mathcal{B}_1/\mathcal{B}_2$.

distinguished. The shorter period fringes (3.8 cm$^{-1}$) correspond to an etalon effect arising from the BaF$_2$ windows of the detectors [25]. The origin of the fringes with the larger period (100 cm$^{-1}$) is unknown. The spikes in the baseline ratio are due to water absorption lines. A slight shift between the two measurements in Fig. 2.17 causes the baseline ratio $\mathscr{B}_1/\mathscr{B}_2$ to be very different from the expected constant 1. This has a detrimental effect on the accuracy of the transmittance [Eq. (2.21)]. Hence, steps must be undertaken to suppress baseline changes between the sample and reference measurement.

### 2.4.4 Characterization of the DFG Spectrometer

The key features of the DFG spectrometer are summarized in Tab. 2.1. With the available pump ($\lambda_p$ = 1064.5 nm) and signal ($\lambda_s$ = 1520–1600 nm) lasers, the idler can be tuned to any wavelength between 2817 cm$^{-1}$ (3.55 µm) and 3144 cm$^{-1}$ (3.18 µm). Quasi phase-matching (QPM) occurs through selection of an appropriate poling period in conjunction with temperature tuning of the crystal. The range from 2900 cm$^{-1}$ to 3144 cm$^{-1}$ can be covered with the $\Lambda$ = 29.9 µm poling period while tuning the PPLN temperature from 40 to 173 °C; the range from 2817 cm$^{-1}$ to 2920 cm$^{-1}$ can be covered with the $\Lambda$ = 29.5 µm poling period while tuning the temperature from 80 to 150 °C. The largest possible tuning range achievable with the PPLN and the current pump laser spans from 2500 cm$^{-1}$ (with $\Lambda$ = 28.5 µm and $\lambda_s$ = 1450 nm) to 3144 cm$^{-1}$ (with $\Lambda$ = 29.9 µm and $\lambda_s$ = 1600 nm). The linewidth of the idler is approx. 150 MHz. The wavenumber step size is given by the tuning characteristics of the signal laser (ECDL). The minimum increment/decrement in signal wavelength is ideally 1 pm with the available device. However, we could only achieve tuning in steps of 10 pm, resulting in idler wavenumber steps of 0.020 cm$^{-1}$ = 600 MHz. The spectral range from 2900 to 3144 cm$^{-1}$ can be scanned in 66.5 min, i.e., with a scan rate of 3.67 cm$^{-1}$/min.

The average power of 150 µW was measured with a thermopile power meter (Ophir 3A, Israel). The peak power was computed from the average power and the duty cycle $\mathscr{D}$ = 3 × 10$^{-5}$ (6 ns pulses at 5 kHz repetition rate).

The signal-to-noise ratio (SNR) of the $\mathfrak{N}$ = 500 times averaged detector signal ratio $Q$ is given in Eq. (2.66). The decay time of the detector pulses generated by the laser pulse is approx. 6 µs, corresponding to $w$ = 300 sample points at 50 MHz sampling rate. The maximum of $R$ in Eq. (2.66) is given by Eq. (2.63) and is obtained for an integration window of width

**Table 2.1** – Key features of the DFG spectrometer.

| Property | Specification | Conditions |
|---|---|---|
| Tuning range | 2817–2920 cm$^{-1}$ (3.42–3.55 µm) | $\Lambda = 29.5\,\mu\text{m}$* |
| | 2900–3144 cm$^{-1}$ (3.18–3.45 µm) | $\Lambda = 29.9\,\mu\text{m}$ |
| Linewidth | 150 MHz = 0.005 cm$^{-1}$ | |
| Wavenumber step size $\Delta\tilde{\nu}$ | 0.002 cm$^{-1}$ | $\Delta\lambda_s = 1\,\text{pm}$† |
| | 0.020 cm$^{-1}$ | $\Delta\lambda_s = 10\,\text{pm}$ |
| Tuning rate | 3.67 cm$^{-1}$/min | $\Delta\tilde{\nu} = 0.03\,\text{cm}^{-1}$ with timetable tuning |
| Power | | 6 ns pulses, 5 kHz repetition rate |
| average $\langle\mathscr{P}\rangle$ | ~ 150 µW | |
| peak $\mathscr{P}_p$ | ~ 5 W | |
| duty cycle $\mathscr{D}$ | $3 \times 10^{-5}$ | |
| Absorption pathlength $\mathscr{L}$ | 35 m | |
| Noise equivalent | | 500 averaged pulses = 450 ms acquisition time per spectral point |
| transmittance $\mathscr{T}_{ne}$ | $1.4 \times 10^{-4} - 1.4 \times 10^{-3}$ | |
| absorbance $\mathscr{A}_{ne}$ | $6.1 \times 10^{-5} - 6.1 \times 10^{-4}$ | |
| absorption coefficient $\alpha_{ne}$‡ | $1.7 \times 10^{-8} - 1.7 \times 10^{-7}\,\text{cm}^{-1}$ | |
| Smallest measurable change in§ | | Deviation of three consecutively measured baselines from a fourth one |
| transmittance $\delta\mathscr{T}$ | $4 \times 10^{-3} - 7 \times 10^{-3}$ | |
| absorbance $\delta\mathscr{A}$ | $1.7 \times 10^{-3} - 3.1 \times 10^{-3}$ | |
| absorption coefficient $\delta\alpha$ | $5.0 \times 10^{-7} - 8.7 \times 10^{-7}\,\text{cm}^{-1}$ | |

*Poling period of the PPLN crystal.
†Minimum increment of the signal laser wavelength.
‡For 35 m pathlength.
§cf. Fig. 2.18 below.

$n \approx 1.26 w \approx 380$ sample points:

$$\text{SNR}_Q = 0.638 \sqrt{\frac{\mathfrak{N} w}{2}} \text{SNR}_p \approx 175 \cdot \text{SNR}_p. \tag{2.67}$$

Hence, the SNR of the detector signal ratio $Q$ with integration and averaging is 175 times larger compared to the SNR of the peak value of the pulse.

The SNR of the peak value $\text{SNR}_p$ varies between 7.5 (at low idler powers) and 200 (at high idler powers), so that $\text{SNR}_Q$ is expected to be between $1.3 \times 10^3$ and $3.5 \times 10^4$. Measured values vary between $10^3$ and $10^4$. The relative noise in the transmittance $\mathcal{T}$ is given by the noise in $Q$ [Eq. (2.19)]:

$$\frac{\delta\mathcal{T}}{\mathcal{T}} = \frac{\delta(Q/\mathcal{B})}{Q/\mathcal{B}} = \sqrt{2}\frac{\delta Q}{Q} = \frac{\sqrt{2}}{\text{SNR}_Q}, \tag{2.68}$$

where we assumed that the relative noise in the baseline $\mathcal{B}$ is the same as in $Q$, and the two are uncorrelated. Hence, the minimum (noise-equivalent) measurable change in transmittance $\mathcal{T}$ from the value 1 is given by

$$\mathcal{T}_{\text{ne}} = \delta\mathcal{T} = \frac{\sqrt{2}}{\text{SNR}_Q} \tag{2.69}$$

and lies between $\sqrt{2} \times 10^{-3}$ and $\sqrt{2} \times 10^{-4}$ (absorbance $\mathcal{A}_{\text{ne}}$: $6.1 \times 10^{-4} - 6.1 \times 10^{-5}$, absorption coefficient $\alpha_{\text{ne}}$: $1.75 \times 10^{-7}\,\text{cm}^{-1} - 1.75 \times 10^{-8}\,\text{cm}^{-1}$ for a total pathlength of 35 m).

In fact, these values are only true if a baseline measurement is not needed, as is the case when measuring narrow spectral features (Fig. 2.4 left). Otherwise, the drift of the baseline $\mathcal{B}$ [Eq. (2.18)] must be taken into account and becomes the limiting factor. Figure 2.18 shows the transmittance error $\delta\mathcal{T}$ introduced by baseline drifts

$$\delta\mathcal{T} = \frac{\mathcal{B}}{\mathcal{B}_0} - 1 \tag{2.70}$$

for three consecutively measured baselines $\mathcal{B}$ from a previously measured one $\mathcal{B}_0$. In Fig. 2.18a the DFG spectrometer was tuned according to the procedure outlined in the previous section (Fig. 2.14). The root mean square values $\delta\mathcal{T}_{\text{rms}}$ [Eq. (2.22)] are indicated in the figure and are between 0.4% and 0.7%. In Fig. 2.18b the same quantities are displayed for three baselines measured with a different tuning procedure. There, the wavelength of the signal laser was tuned in steps of 10 pm from 1540 to 1600 nm and the PPLN crystal temperature, computed by inverting Eq. (2.53), was continuously matched to the current signal wavelength. Consequently, the duration is not equal for all measurements and the average duration is longer (4.5 hours) compared to the timetable tuning procedure (with the same number of spectral points). Additionally, the reproducibility of the spectra is worse, as can be gathered from the larger values of $\delta\mathcal{T}_{\text{rms}} \gtrsim 1.3\%$. The values found for $\delta\mathcal{T}_{\text{rms}}$ with the timetable method (0.4–0.7%) correspond to absorbances $\delta\mathcal{A}$ of $1.7 \times 10^{-3} - 3.1 \times 10^{-3}$ and absorption coefficients $\delta\alpha$ of $5.0 \times 10^{-7} - 8.7 \times 10^{-7}\,\text{cm}^{-1}$ (for 35 m pathlength), between a factor 3 and

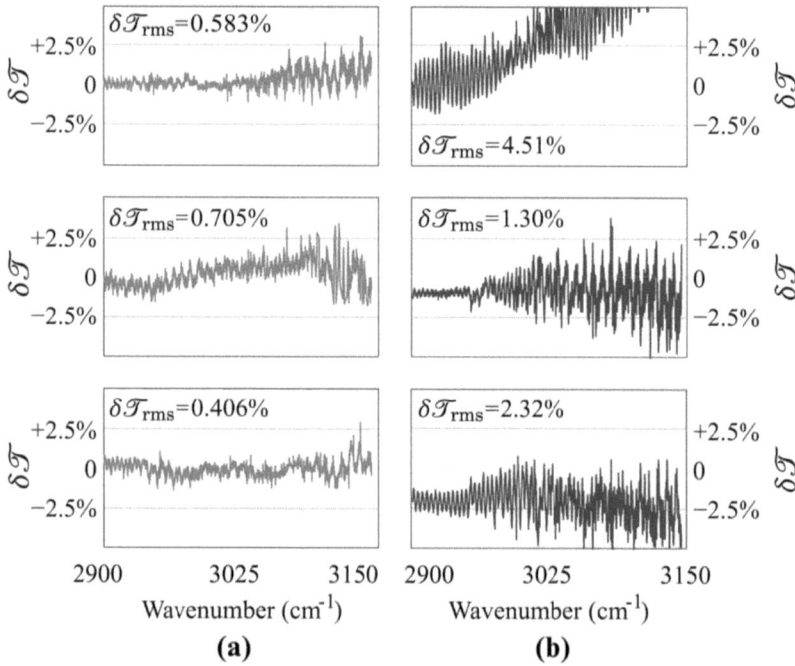

**Figure 2.18** – (a) Transmittance error $\delta\mathcal{T}$ due to baseline drifts for three consecutive baseline measurements $\mathcal{B}$ compared to an initial measurement $\mathcal{B}_0$ [Eq. (2.70)] when the spectrometer is tuned with the timetable method (duration of one scan = 67 min). (b) Same as (a), but the baselines were measured by selecting the signal laser wavelength first (in steps of 10 pm), and matching the PPLN crystal temperature (duration of one scan = 4.5 hours).

50 worse than in the (baseline-drift-free) noise-limited case [Eq. 2.69]. It should be noted that these values are valid only for a single pure substance measured against a non-absorbing background, i.e., if no other substance interferes with the detection of a target compound.

The effective detection bandwidth is difficult to compute due to the pulsed nature of the signals. To compute the transmittance at one specific wavelength, the detector signal ratio $Q$ [Eq. (2.64)] must be measured twice: once for the sample spectrum and once for the reference spectrum. Since the acquisition of one set of signal pulses takes about 150 ms (Fig. 2.14c), and taking into account that two measurements are needed, we can assume a detection bandwidth of about $(2 \cdot 150 \text{ ms})^{-1} = 3.3$ Hz. With cavity ring-down (CRD) typical smallest measurable absorption coefficients are of the order of $10^{-9}$–$10^{-10}$ cm$^{-1}$Hz$^{-1/2}$ [100–102]. Similar values can be achieved with photoacoustics (PA) [103] and wavelength modulation spectroscopy [103–105]. Both CRD and PA require sufficient laser power to be effective. Moreover, CRD requires high-reflectivity mirrors over the entire tuning range. Highest sensitivity, but at the price of extreme complexity, is provided by noise-immune cavity enhanced optical heterodyne molecular spectroscopy (NICE-OHMS), with smallest measurable absorbances in the range of $10^{-8}$–$10^{-13}$ [106].

## 2.5 Distributed Feedback Laser Diode Spectrometer

### 2.5.1 Setup

As outlined in Sec. 2.2, we have also used a near-IR diode laser spectrometer for our studies. The distributed feedback (DFB) laser spectrometer is shown in Fig. 2.19. Two DFB laser diodes (nanoplus, Germany) are used as laser sources. One (diode A) emits at 2433.1 nm (4110 cm$^{-1}$) and the other (diode B) at 2323.6 nm (4303.6 cm$^{-1}$). These two wavelengths correspond to hydrogen fluoride (HF) and carbon monoxide (CO) absorptions, respectively. Both diodes are housed in thermoelectrically (TE) cooled laser diode mounts (Thorlabs, Germany, TCLDM9). The two lasers cannot be

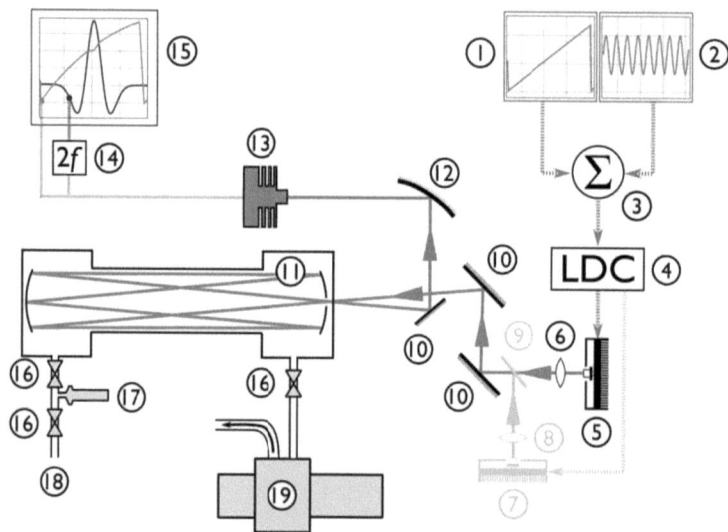

**Figure 2.19** – Schematic drawing of the DFB laser spectrometer. (1) voltage ramp generator; (2) voltage sine generator; (3) adder; (4) laser diode current driver; (5) thermoelectrically (TE) cooled laser diode mount and DFB laser diode A ($\lambda$ = 2433.1 nm); (6) collimating CaF$_2$ $f$ = 12.7 mm lens; (7) TE cooled laser diode mount and DFB laser diode B ($\lambda$ = 2323.6 nm); (8) collimating CaF$_2$ $f$ = 12.7 mm lens; (9) flipping mirror; (10) silver-coated mirror; (11) multipass gas cell (MPC) with CaF$_2$ window; (12) off-axis parabolic silver-coated mirror ($f$ = 10 cm); (13) TE cooled HgCdZnTe detector; (14) lock-in amplifier (second harmonic detection); (15) oscilloscope or acquisition card; (16) valve; (17) pressure gauge; (18) inlet for gas samples; (19) rotary vacuum pump.

used simultaneously, but a flipping mirror allows to quickly switch from one diode to the other. Two function generators produce a voltage ramp and sine (frequency $f$), which are added and then fed to a laser diode current driver (Thorlabs, Germany, model LDC 500). The divergent laser beams are collimated by $f = 12.7$ mm $CaF_2$ lenses placed immediately after the diode mounts, and led into the multipass cell (MPC, New Focus, U.S.A., model 5611 with $CaF_2$ window) via two silver-coated mirrors. An off-axis parabolic silver-coated mirror ($f = 10$ cm) focuses the transmitted beam onto a TE cooled preamplified HgCdZnTe detector (Vigo Systems SA, Poland, model PV-2TE-4/VPDC-10i). The peak responsivity of the detector is at $4 \mu m$, but it is still sufficiently high at $2.3$–$2.4 \mu m$ with the available powers of the diodes (2–3 mW). The voltage signal generated by the detector is split in two: one part is demodulated at the frequency $2f$ with a lock-in amplifier (Stanford Research Systems, U.S.A., model SR830) and then acquired with an oscilloscope (Tektronix, U.S.A., model TDS 644A), and the other is acquired directly with the same oscilloscope or with an acquisition card (Gage, U.S.A., model CS14100). A rotary vacuum pump (Alcatel, France) is used to evacuate the multipass cell. The pressure inside the cell can be monitored with a pressure gauge (Balzers, model APG 010). Both the two laser diode mounts and the MPC can be "shaken" with piezo transducers (PZT). The mirror separation distance of the MPC is altered by a few microns when the PZT is active. This is needed to remove optical fringes (see Sec. 2.5.4 and Fig. 2.23 below).

The key features of the spectrometer are described later in Sec. 2.5.4 and in Tab. 2.2.

## 2.5.2 Data Acquisition and Evaluation

The DFB laser spectrometer can be operated both in wavelength modulation (WM) and in direct transmission mode. The only difference is that in direct transmission mode the sinusoidal laser wavelength modulation is switched off and the lock-in amplifier is not used.

In WM mode the signal is sent through a lock-in amplifier that demodulates it at twice the modulation frequency $f$. The output of the lock-in amplifier is then acquired by an 8-bit digital sampling oscilloscope. This relatively low resolution is enhanced by measuring each spectrum several times at a repetition rate of about 3.3 Hz and then averaging them. The oscilloscope does the averaging and is triggered by the voltage ramp. At the end of the measurements, the averaged spectrum is transferred from the oscilloscope to the computer with a home-written LabVIEW (National Instruments, U.S.A.) program. The wavelength axis is calibrated by measuring a CO (diode B) or a CH$_4$ (diode A) sample. The exact absorption line positions are extracted from the HITRAN database [107]. The peak-to-peak or zero-to-peak (see Fig. 2.5 for the definitions) values of each identified line are then used to compute concentrations (see Sec. 2.5.3).

In direct transmission mode the detector signal is acquired directly. Since a better resolution is needed (because of the strong baseline), a 14-bit ADC card is used and the signal is oversampled at 100 MHz and then decimated to enhance resolution. The scan repetition rate is increased to about 1 kHz to achieve a lower noise level (see Sec. 2.3.2). The baseline is fitted with a polynomial and divided from the measured signal. This procedure provides the true transmittance so that a calibration is not necessary.

### 2.5.3 Calibration

Figure 2.20 shows the computed spectra of water, methane, carbon monoxide and hydrogen fluoride near 4110 cm$^{-1}$ and 4303.6 cm$^{-1}$ for a total pres-

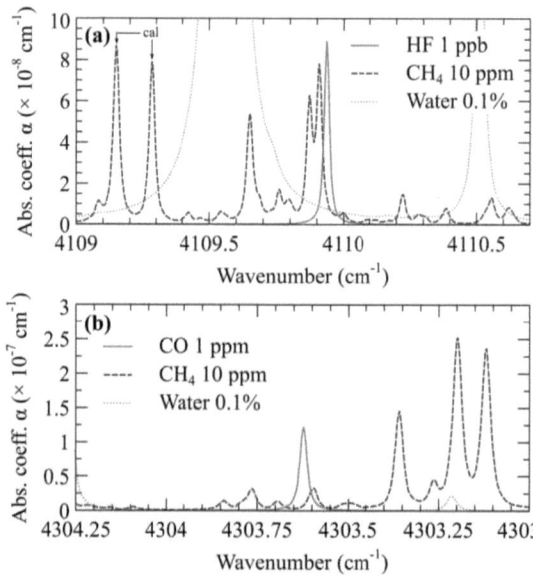

**Figure 2.20** – **(a)** Computed spectra of hydrogen fluoride (1 ppb), methane (10 ppm) and water vapor (0.1%) near 4110 cm$^{-1}$ at 100 mbar. The two methane absorption lines indicated with *cal* are used for calibration. **(b)** Computed spectra of carbon monoxide (1 ppm), methane (10 ppm) and water vapor (0.1%) near 4303.6 cm$^{-1}$ at 100 mbar.

sure $p = 100\,\text{mbar}$. Since hydrogen fluoride is delicate to handle, toxic and corrosive, and producing samples with concentrations around 1 ppb is challenging, diode A was calibrated with methane instead. Because the two methane lines near the hydrogen fluoride line partially overlap, two different lines were chosen for the calibration. Diode B was calibrated with the actual carbon monoxide line at $4303.6\,\text{cm}^{-1}$ (Fig. 2.20b).

The total absorption pathlength $\mathscr{L}$ of the spectrometer was determined by measuring one carbon monoxide (300 ppm buffered in nitrogen) and one methane (1000 ppm buffered in nitrogen) absorption line at a total pressure of 100 mbar with the two DFB laser diodes in direct transmission mode. The absorbance spectrum was also computed by using the HITRAN data [107] at the given concentrations and fitted to the measured spectra by varying the pathlength. The result is shown in Fig. 2.21. With both diodes the total

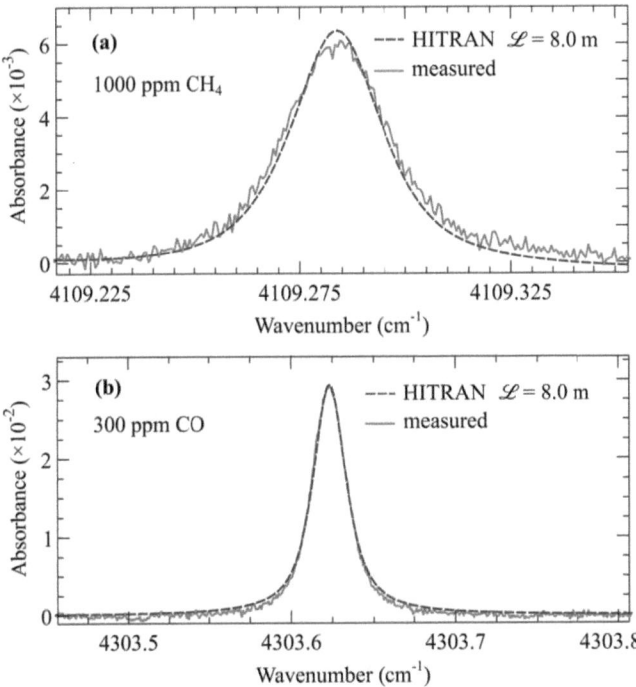

**Figure 2.21** – (a) Methane (1000 ppm in nitrogen) absorption line measured with diode A and fitted spectrum (data from HITRAN) for a pathlength of 8.0 m. (b) Carbon monoxide (300 ppm in nitrogen) absorption line measured with diode B and fitted spectrum (data from HITRAN) for a pathlength of 8.0 m.

pathlength $\mathscr{L}$ amounts to 8.0±0.1m. The deviation from the expected value of 18 or 36 m for this cell could be caused by the bad beam quality.

In Sec. 2.3.3 it was shown that the 2f amplitude (f = modulation frequency of the laser wavelength) of the transmitted power at the center of the absorption line is proportional to $1-\mathscr{T}_0$, where $\mathscr{T}_0$ is the transmittance at the center of the absorption line [Eqs. (2.37) and (2.38)]:

$$\frac{\mathscr{S}_w}{\mathscr{P}} = \mathfrak{F}_2(\mathscr{T}_0) = k(1-\mathscr{T}_0), \tag{2.71}$$

where $\mathscr{P}$ is the power incident on the detector. From Eq. (2.40) we expect $k = 0.438$ for a purely Doppler-broadened absorption line and with modulation index $m = 2.1$. For both diodes a set of calibration measurements with methane and carbon monoxide samples at a pressure of $p = 100$ mbar and at different known concentrations was carried out. The normalized 2f signal amplitudes $\mathscr{S}_w/\mathscr{P}$ ($= \mathfrak{F}_2$) at the absorption line position versus $1-\mathscr{T}_0$ is shown in Fig. 2.22. The values of $\mathscr{T}_0$ were derived from the HITRAN database [107]. For the diode A (Fig. 2.22a), two methane lines with slightly different absorbances (indicated with *cal* in Fig. 2.20) were measured for five different concentrations (this is why the measured points ap-

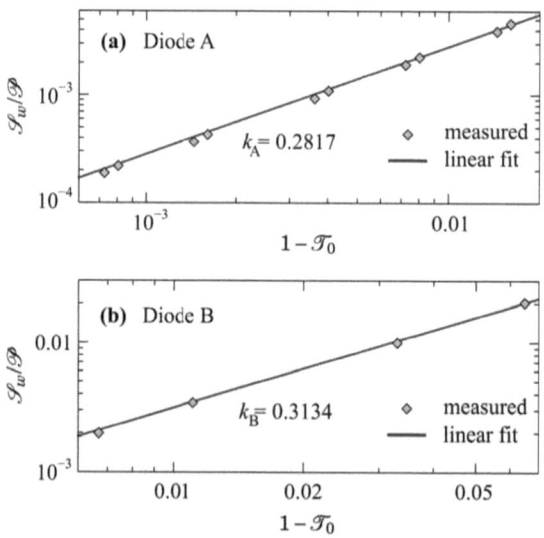

**Figure 2.22** – (a) Normalized 2f signal amplitude of two methane absorption lines as a function of $1-\mathscr{T}_0$ ($\mathscr{T}_0$ = transmittance at the center of the absorption line) measured with diode A. (b) Normalized 2f signal amplitude of one carbon monoxide absorption lines as a function of $1-\mathscr{T}_0$ measured with diode B.

pear paired). The linear fit yields $k_A = 0.2817$. The expected value 0.438 was computed assuming a perfect Gaussian lineshape, modulation index $m = 2.1$ and no laser intensity modulation. Neither of these conditions are fulfilled exactly, which could explain the discrepancy. For the diode B (Fig. 2.22b), a single absorption line at four different concentrations of carbon monoxide was measured. The linear fit yields $k_B = 0.3134$. With both diodes the linearity is good, as shown by the linear fits in Fig. 2.22.

### 2.5.4 Characterization of the Laser Diode Spectrometer

The key features of the DFB spectrometer are summarized in Tab. 2.2. With the central laser wavelength kept at a constant value, the 2f amplitude $\mathscr{S}_w$ [Eq. (2.37)] and the power $\mathscr{P}$ incident onto the detector have been measured over two minutes. The lock-in time constant was set to 1 ms and the low-pass filter roll-off to 24 dB/oct. This corresponds to an equivalent noise bandwidth of 78 Hz. Additionally, 200 measurements were averaged, so that the final detection bandwidth was $B = 0.39$ Hz. For the diode A, the absolute noise of the normalized signal, $\mathscr{S}_w/\mathscr{P}$, was $s_A = 2.7 \times 10^{-7}$ which leads to a noise equivalent transmittance of

$$\mathscr{T}_{\text{ne}} = \frac{s_A}{k_A} \approx 9.7 \times 10^{-7}. \qquad (2.72)$$

Equation (2.71) and $k_A = 0.2817$ were used in Eq. (2.72). The noise equivalent absorbance is $\mathscr{A}_{\text{ne}} = 4.2 \times 10^{-7}$ and the noise equivalent absorption coefficient is $\alpha_{\text{ne}} = 5.3 \times 10^{-10}$ cm$^{-1}$ for a pathlength of 8 m. The 1-Hz bandwidth-normalized values, defined via

$$\mathscr{A}_{\text{ne}} = \mathscr{A}_{1\,\text{Hz}} \cdot \sqrt{B} \qquad (2.73)$$

and similar for the other parameters, are given in Tab. 2.2. For the diode B, the noise values were slightly higher, $s_B = 3.1 \times 10^{-7}$, which corresponds to

$$\mathscr{T}_{\text{ne}} = \frac{s_B}{k_B} = 9.8 \times 10^{-7}. \qquad (2.74)$$

In direct transmission mode similar values are achieved.

Unfortunately there is another phenomenon limiting the sensitivity of the spectrometer. Figure 2.23a shows a carbon monoxide absorption line measured with diode B with wavelength modulation and 2f detection at a concentration of 500 ppb. Strong interference fringes mask the absorption line and make a detection impossible. By vibrating both the laser diode mount and the multipass cell (by varying the mirror separation distance)

**Table 2.2** – Key features of the DFB spectrometer in wavelength modulation mode.

| Property | Specification | | Conditions |
|---|---|---|---|
| | Diode A (HF) | Diode B (CO) | |
| Wavelength | 2433.1 nm (4110 cm$^{-1}$) | 2323.6 nm (4303.6 cm$^{-1}$) | |
| Linewidth | < 3 MHz | < 3 MHz | specified by manufacturer |
| Power | 2 mW | 3 mW | continuous wave |
| Absorption pathlength $\mathscr{L}$ | 8.0 m | 8.0 m | |
| Noise equivalent | | | |
| transmittance $\mathscr{T}_{1\,Hz}$ | $1.6 \times 10^{-6}$ Hz$^{-1/2}$ | $1.6 \times 10^{-6}$ Hz$^{-1/2}$ | |
| absorbance $\mathscr{A}_{1\,Hz}$ | $6.8 \times 10^{-7}$ Hz$^{-1/2}$ | $6.8 \times 10^{-7}$ Hz$^{-1/2}$ | |
| absorption coefficient $\alpha_{1\,Hz}$ * | $8.5 \times 10^{-10}$ cm$^{-1}$ Hz$^{-1/2}$ | $8.5 \times 10^{-10}$ cm$^{-1}$ Hz$^{-1/2}$ | |
| concentration $c_{1\,Hz}$ | 10 ppt Hz$^{-1/2}$ | 7 ppb Hz$^{-1/2}$ | |
| Smallest measurable change in | | | |
| transmittance $\delta\mathscr{T}$ | $1.8 \times 10^{-5}$ | $5.7 \times 10^{-5}$ | |
| absorbance $\delta\mathscr{A}$ | $8.0 \times 10^{-6}$ | $2.5 \times 10^{-5}$ | |
| absorption coefficient $\delta\alpha$ | $1.0 \times 10^{-8}$ cm$^{-1}$ | $3.1 \times 10^{-8}$ cm$^{-1}$ | For a total measurement time of 2 min. |
| concentration $\delta c$ | 110 ppt | 250 ppb | |

*For 8 m pathlength.

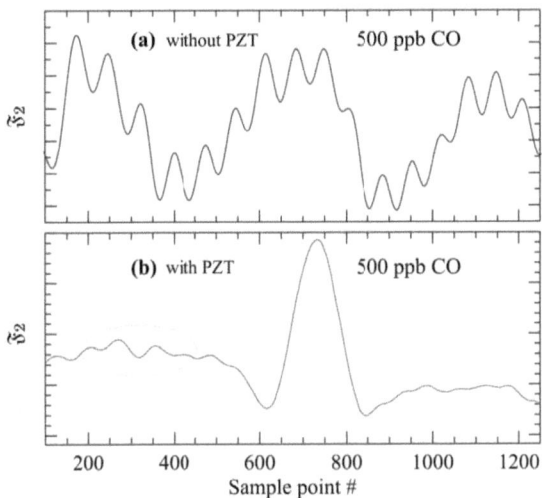

**Figure 2.23** – (a) Absorption line of CO (at 500 ppb) measured with wavelength modulation (second harmonic amplitude, $\mathfrak{F}_2$) hidden under interference fringes. (b) Absorption line of CO (500 ppb) measured with wavelength modulation and piezo transducer dithering. Some fringes are left over (circled region).

at low (10–30 Hz) frequency and by only a few microns, the interference pattern continuously shifts left and right. By averaging several spectra (here, 200) the fringes almost completely cancel out (Fig. 2.23b). The effectiveness of this cancellation is not reproducible from one series of measurements to the next one, but appears to be, at least in part, random. For measurement times up to about 2 min (detection bandwidth $B = 0.39\,\text{Hz}$), both the fringe and noise amplitude decrease. For longer measurement times, the noise further decreases like $1/\sqrt{B}$, but the amplitude of the residual fringes remains constant, so that there is no further gain in sensitivity. Figure 2.24 shows the spectrum of the evacuated multipass cell measured with diode A. The scale on the vertical axis is given by

$$\alpha = \frac{\mathscr{A}}{\mathscr{L}} \approx \frac{1-\mathscr{T}}{\mathscr{L}\ln 10} = \frac{\mathscr{S}_w}{\mathscr{P}}\frac{1}{k_A \mathscr{L}\ln 10}. \tag{2.75}$$

Equation (2.16) has been used in the second and Eq. (2.71) in the third equality. The smallest measurable change in absorption coefficient can be estimated from Fig 2.24 to be approximately $\delta\alpha = 1 \times 10^{-8}\,\text{cm}^{-1}$. For diode B a similar measurement leads to a smallest measurable absorption coefficient of $\delta\alpha = 3.1 \times 10^{-8}\,\text{cm}^{-1}$.

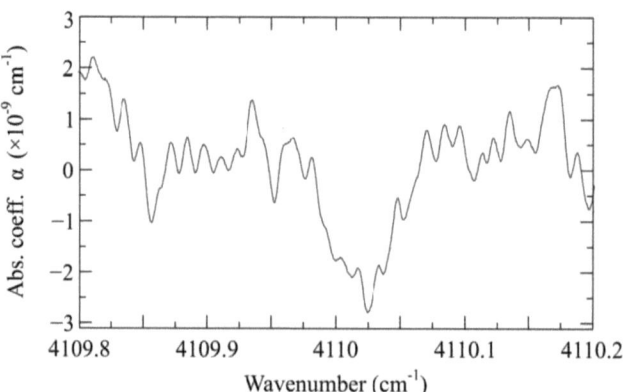

**Figure 2.24** – Spectrum of the evacuated multipass cell measured with wavelength modulation (second harmonic detection) with diode A.

## 2.6 External Cavity Quantum Cascade Laser Spectrometer

### 2.6.1 Setup

As mentioned in Sec. 2.2, we used an ECQCL as an alternative laser source for some of our analytical studies. The external cavity quantum cascade laser (ECQCL) spectrometer is shown in Fig. 2.25. The laser source is an ECQCL (Daylight Solutions, U.S.A.) tunable from 1005 to 1100 cm$^{-1}$ in steps of 0.9 cm$^{-1}$ with up to 200 mW of output power. It can be operated in both continuous wave and pulsed mode. The linewidth is specified by the manufacturer as < 100 MHz. A mechanical gold-coated chopper (New Focus, U.S.A.), aside from modulating the intensity of the beam, acts as a beam splitter, alternately transmitting the beam into the multipass cell

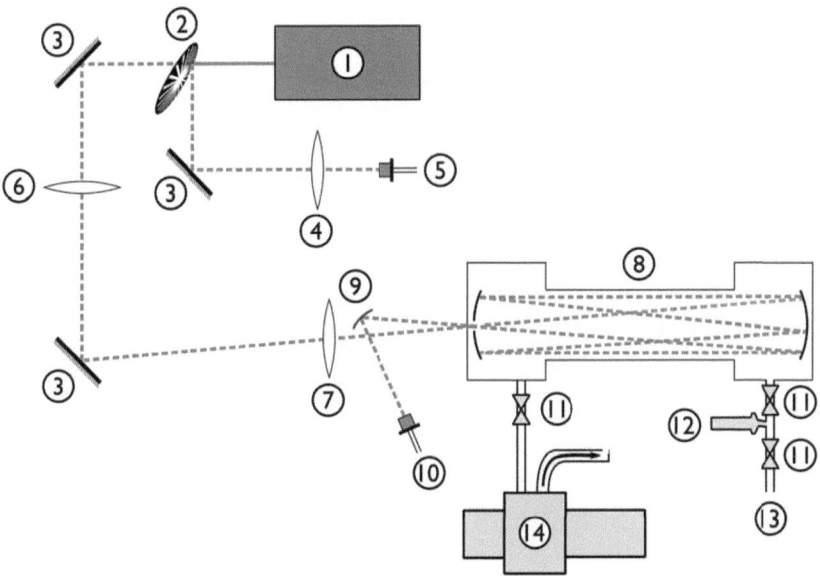

**Figure 2.25** – Schematic representation of the ECQCL spectrometer. (1) EC-QCL (1005–1100 cm$^{-1}$); (2) chopper; (3) silver-coated mirrors; (4) AR-coated $f$ = 75 mm ZnSe lens; (5) reference pyroelectric detector; (6) AR-coated $f$ = 2000 mm ZnSe lens; (7) AR-coated $f$ = 200 mm ZnSe lens; (8) multipass cell ($\mathscr{L}$ = 36 m); (9) concave $f$ = 100 mm silver-coated mirror; (10) transmission pyroelectric detector; (11) valves; (12) pressure gauge; (13) inlet for gas samples; (14) vacuum pump.

(New Focus, U.S.A., model 5611 with ZnSe window) and reflecting it into the reference detector. Both detectors are room-temperature windowless pyroelectric detectors (Eltec Instruments, U.S.A., model 420). A $f = 75\,\text{mm}$ ZnSe lens focuses the beam reflected off the chopper onto the reference detector. A weak $f = 2\,\text{m}$ ZnSe lens is used to collimate the slightly divergent beam before a stronger $f = 200\,\text{mm}$ ZnSe lens focuses it into the center of the multipass cell. The total absorption pathlength is 36 m, corresponding to 182 passes through the multipass cell. After the beam leaves the cell, a concave metallic mirror ($f = 100\,\text{mm}$) reflects it onto the transmission detector. Two lock-in amplifiers (Stanford Research Systems, U.S.A., model SR830) demodulate the signal generated by the detectors at the modulation frequency given by the chopper.

### 2.6.2 Data Acquisition and Evaluation

A spectrum is measured by scanning the laser wavelength from 1005 to $1100\,\text{cm}^{-1}$ during 6 seconds and measuring the detector signal ratio

$$Q \equiv \frac{D_T}{D_R}, \tag{2.76}$$

where $D_R$ and $D_T$ are the reference and transmission signal amplitudes, respectively. To prevent loss of resolution, the lock-in time constant should not be larger than 30 ms. To further decrease the noise, several measurements (typically 35) are averaged. Because this technique is not baseline-free, a reference measurement $\mathscr{B}$ is needed. The transmittance $\mathscr{T}$ is then given by Eq. (2.19).

### 2.6.3 Characterization of the External Cavity Quantum Cascade Laser Spectrometer

The key features of the ECQCL spectrometer are given in Tab. 2.3. The noise equivalent power of the detectors was determined to be approximately $10^{-7}\,\text{W}\cdot\text{Hz}^{-1/2}$ at 1 kHz modulation frequency. The total relative noise in the detector signal ratio $Q$ [Eq. (2.76)], measured by keeping the wavelength constant, had a value of $1/\text{SNR}_Q = 4.4 \times 10^{-3}\,\text{Hz}^{-1/2}$. The minimum measurable change in transmittance is given by Eq. (2.69) and is $\mathscr{T}_{1\,\text{Hz}} = 6.2 \times 10^{-3}\,\text{Hz}^{-1/2}$ (absorbance $\mathscr{A}_{1\,\text{Hz}} = 2.7 \times 10^{-3}\,\text{Hz}^{-1/2}$, absorption coefficient $\alpha_{1\,\text{Hz}} = 7.5 \times 10^{-7}\,\text{cm}^{-1}\cdot\text{Hz}^{-1/2}$ for 36 m pathlength).

However, as in the case of the DFG spectrometer (see Sec. 2.4.4), the sensitivity is limited by the reproducibility of the measurements. Since a single measurement only lasts 6 s, several measurements can be averaged

**Table 2.3** – Key features of the ECQCL spectrometer.

| Property | Specification | Conditions |
|---|---|---|
| Tuning range | 1005–1100 cm$^{-1}$ (9.09–9.95 µm) | |
| Linewidth | < 100 MHz | specified by manufacturer |
| Power | up to 200 mW | continuous wave |
| Absorption pathlength $\mathscr{L}$ | 36 m | |
| Noise equivalent | | |
| transmittance $\mathscr{T}_{1\,Hz}$ | 6.2 × 10$^{-3}$ Hz$^{-1/2}$ | |
| absorbance $\mathscr{A}_{1\,Hz}$ | 2.7 × 10$^{-3}$ Hz$^{-1/2}$ | |
| absorption coefficient $\alpha_{1\,Hz}$* | 7.5 × 10$^{-7}$ cm$^{-1}$Hz$^{-1/2}$ | |
| Smallest measurable change in | | Averaging 35 individual scans, detection bandwidth $B = 0.074$ Hz |
| transmittance $\delta\mathscr{T}$ | 2.1 × 10$^{-3}$ | |
| absorbance $\delta\mathscr{A}$ | 9.1 × 10$^{-4}$ | |
| absorption coefficient $\delta\alpha$ | 2.5 × 10$^{-7}$ cm$^{-1}$ | |

*For 36 m pathlength.

in a reasonable amount of time. The optimum number of averages is a compromise between noise suppression and measurement duration. If more scans are averaged the noise can be gradually reduced, but the duration increases and with it the baseline drifts between the sample and reference measurement. We consecutively measured $N_m$ spectra of the multipass cell filled with nitrogen and then grouped them in $N_g \equiv \lfloor N_m/n \rfloor$ groups of size $n$, so that the first $n$ spectra were in group one, the second set of $n$ spectra were in group two, and so on. The spectra in each group were averaged, and the average spectrum of group one was divided by the average spectrum of group two, and so on. Of each division the root mean square (r.m.s.) deviation from the expected value 1 was taken and then all r.m.s. values were averaged:

$$s(n) \equiv \left\langle \sqrt{\left\langle \left(\frac{X_{2k-1}}{X_{2k}} - 1\right)^2 \right\rangle_\lambda} \right\rangle_p, \quad (2.77)$$

where $X_j$ is the average spectrum of group $j$, the innermost average $\langle \rangle$ is performed over all wavelengths ($\lambda$), and the outermost is performed over all $N_g/2$ pairs ($p$) of groups. In Eq. (2.77), $s(n)$ is the r.m.s. value of the baseline drift-induced error between two $n$-times averaged spectra. Notice that the term under the square root corresponds to $\langle(\delta\mathcal{T})^2\rangle$ [Eqs. (2.21) and (2.22)]. The only difference between Eq. (2.22) and (2.77) is that with the ECQCL the measurements are much quicker (6s versus 66min for the DFG spectrometer), so that we can choose how often to repeat the scan. The idea behind the definition of $s$ [Eq. (2.77)] is to introduce a quantity analogous to the Allan variance [108] but for a spectrum instead of a single variable. The measured $s(n)$ is shown in Fig. 2.26a. The minimum of $s(n)$ is at $n = 35$ with a value of $\delta\mathcal{T} = 2.1 \times 10^{-3}$. This error in transmittance corresponds to an absorbance error of $\delta\mathcal{A} = 9.1 \times 10^{-4}$ ($\delta\alpha = 2.5 \times 10^{-7}\,\text{cm}^{-1}$ for 36m absorption pathlength). The time constant of the lock-in amplifiers was set to 30ms, so that the total detection bandwidth (including averaging) was $B = 0.074\,\text{Hz}$. For narrower detection bandwidths (more averaging), there is no further gain in sensitivity, meaning that the additional noise reduction is compensated by larger baseline drifts. The dashed line in Fig. 2.26a illustrates the expected profile for $s$ if only white noise were present ($s \sim 1/\sqrt{n}$). Figure

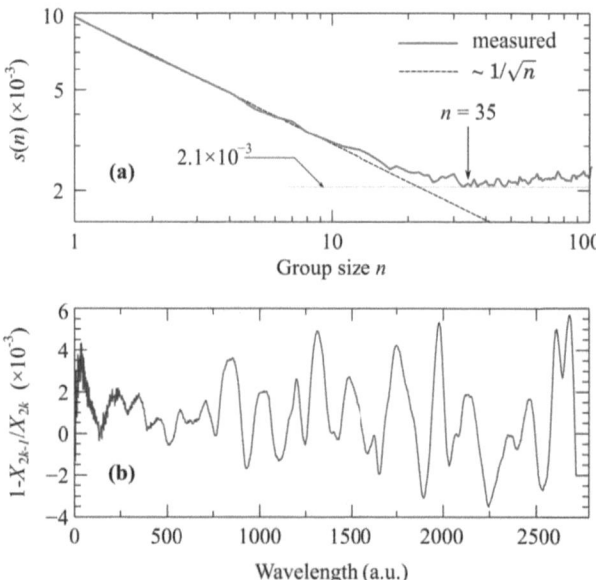

**Figure 2.26** – **(a)** Root mean square transmittance error $s(n)$ [Eq. (2.77)] between two successively measured $n$ times averaged spectra as a function of $n$. **(b)** Transmittance error caused by baseline drifts between two series of 35 averaged measurements.

2.26b shows the transmittance error between two series of 35-times averaged spectra. The r.m.s. value is $2.1 \times 10^{-3}$, but can be as large as $6 \times 10^{-3}$ at some wavelengths.

### 2.6.4 Preliminary Measurements

The finite wavenumber step size of the ECQCL ($0.9\,\mathrm{cm}^{-1}$) is not well suited to measure spectra that contain absorption lines with linewidths $\ll 1\,\mathrm{cm}^{-1}$. Since our final goal is to analyze smoke samples that may contain some trace compounds in a large background of carbon dioxide, we performed a few preliminary measurements with the ECQCL spectrometer to ascertain whether it is at all suitable for measuring gas samples with large concentrations of carbon dioxide.

We measured a carbon dioxide sample twice, and computed the ratio between the two spectra (Fig. 2.27). The absorption is saturated at several wavelengths and, due to the wavenumber step size of $0.9\,\mathrm{cm}^{-1}$, the

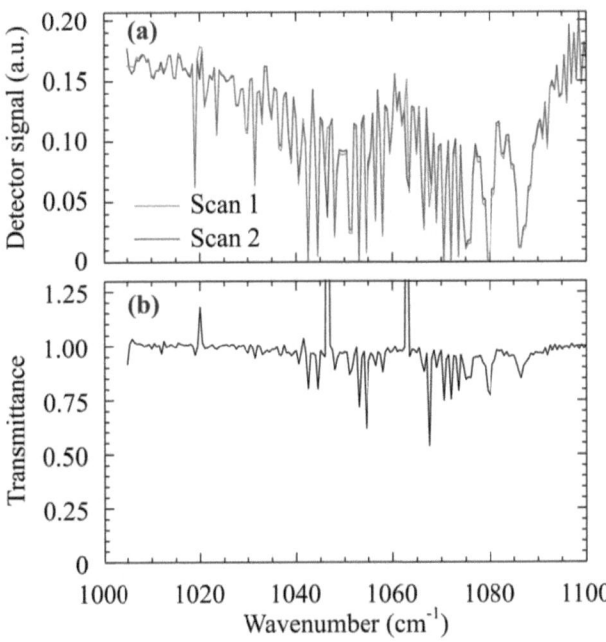

**Figure 2.27** – **(a)** Two spectra of a carbon dioxide sample ($p$ = 960 mbar, $T$ = room temperature). **(b)** Computed transmittance if scan 1 is considered the sample and scan 2 the reference measurement (background).

lines appear very jagged (Fig. 2.27a). Furthermore, the baseline subtraction is inefficient, probably because the two spectra were not sampled at the same exact wavelengths, or because the pressure (and thus the linewidths) changed slightly between the two measurements (Fig. 2.27b).

We conclude that this spectrometer is inapplicable for measurements on gas mixtures that contain close to 100% carbon dioxide. In a non-absorbing buffer gas such as nitrogen, the wavenumber step size of $0.9\,\text{cm}^{-1}$ is problematic for spectra with narrow absorption lines, but is tolerable if only broad absorption bands are present (Fig. 2.28). In Fig. 2.28a the spectrum of about 25 ppm methanol buffered in nitrogen is shown. While the central feature (Q–branch) is clearly defined, some distortion in the P and R–branches can be noticed. The spectrum of m-xylene (Fig. 2.28b), which only manifests broad absorption bands, can be measured with little distortion.

The ECQCL spectrometer was not used for the study of surgical smoke samples (see Ch. 5) because of their high carbon dioxide concentration.

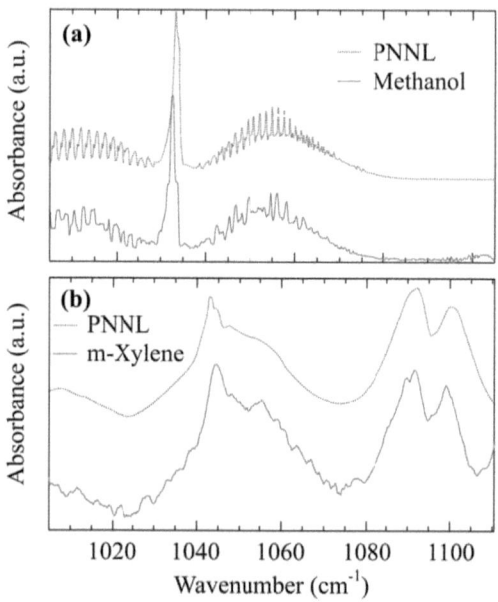

**Figure 2.28** – (a) Measured spectrum of methanol (about 25 ppm) buffered in nitrogen and methanol spectrum from the PNNL database [109] ($p = 960$ mbar, $T$ = room temperature). (b) Measured spectrum of m-xylene (about 115 ppm, $p = 960$ mbar, $T$ = room temperature) buffered in nitrogen and m-xylene spectrum from the PNNL database.

# Chapter 3

# Analysis of Infrared Spectra

In this chapter we present an algorithm for the identification and quantification of components of a gas mixture based on the comparison between the sample's infrared absorption spectrum and a library of infrared spectra of pure substances.

## 3.1 Spectra of Multicomponent Gas Mixtures

The definitions of transmittance $\mathcal{T}$, absorbance $\mathcal{A}$ and concentration $c$ have been given previously [Eqs. (2.9), (2.12) and (2.13)]. If a gas mixture contains several absorbers $a$ all at low concentration $c_a \ll 1$, then interactions between them can be neglected, and the transmission spectrum of the mixture is the product of the transmission spectra of every single component of the mixture:

$$\mathcal{T} = \prod_a \mathcal{T}_a. \tag{3.1}$$

The absorbance $\mathcal{A}$ is then:

$$\mathcal{A} \equiv -\log \mathcal{T} = -\sum_a \log \mathcal{T}_a = \sum_a \mathcal{A}_a \propto \mathscr{L} \sum_a c_a \sigma_a(\lambda), \tag{3.2}$$

where Eq. (2.13) has been used in the last step: $\sigma_a$ is the absorption cross section of absorber $a$ and $\mathscr{L}$ is the total pathlength. Hence, the absorbance spectrum of a gas mixture is the sum of the absorbance spectra of its pure components. The quantitative analysis of a gas mixture therefore consists in identifying the components $a$ of the measured spectrum, and then determining the concentrations $c_a$ that yield the closest match between the measured spectrum and the absorbance computed in Eq. (3.2), where all $\sigma_a$ are taken from a spectral library.

## 3.2 Formulation of the Problem

A spectrum is a sequence of absorbance values (or other arbitrary units proportional to the absorbance) recorded at fixed wavelengths. It can be regarded as a row vector

$$\boldsymbol{x} \equiv (x_1 \; x_2 \; \ldots \; x_p), \tag{3.3}$$

with $p$ denoting the number of wavelength points. Assume that the spectra of $s$ (with $s < p$) pure substances sampled at the same $p$ wavelengths as $\boldsymbol{x}$ are given in a $s \times p$ matrix $\boldsymbol{D}$. Since the absorbance spectrum $\boldsymbol{x}$ is a linear superposition of the spectra of the pure substances [Eq. (3.2)] we can write

$$\boldsymbol{x} = \boldsymbol{cD} + \boldsymbol{\epsilon}, \tag{3.4}$$

where $\boldsymbol{\epsilon}$ is a $1 \times p$ vector of residuals and $\boldsymbol{c}$ is a $1 \times s$ vector of concentrations. The residual vector $\boldsymbol{\epsilon}$ is necessary because $\boldsymbol{x}$ might point out of the $s$-dimensional space spanned by the rows of $\boldsymbol{D}$. The solution $\boldsymbol{c}$ of Eq. (3.4) which minimizes the sum of the squares of the residuals, $\boldsymbol{\epsilon\epsilon}^\mathrm{T}$, is[1]

$$\boldsymbol{c} = \boldsymbol{xD}^\mathrm{T} \left( \boldsymbol{DD}^\mathrm{T} \right)^{-1}. \tag{3.5}$$

This procedure, although very simple and fast – the matrix inversion $(\boldsymbol{DD}^\mathrm{T})^{-1}$ has to be computed only once – has two drawbacks. First, Eq. (3.5) produces positive and negative concentrations, but negative values don't have a physical meaning. Second, if some spectra in $\boldsymbol{D}$ are very similar (nearly collinear), or are almost linear combinations of each other, then $\boldsymbol{DD}^\mathrm{T}$ may be ill-conditioned, and the inverse $(\boldsymbol{DD}^\mathrm{T})^{-1}$ becomes inaccurate. In such a case it becomes necessary to remove one or more rows of $\boldsymbol{D}$, and one would like to only discard redundant spectra and thus lose a minimum amount of information.

---

[1] This can be shown by setting the derivative of $\boldsymbol{\epsilon\epsilon}^\mathrm{T}$ with respect to $c_k$ to zero for all $k$.

## 3.3 Principal Component Analysis

Principal component analysis (PCA) is often used when one wishes to describe a highly multidimensional dataset with only a few characteristic values. One field of application is *facial recognition*, where digital images of faces ($> 10^5$ pixels) can be reduced to only 100–150 values with a technique based on PCA, known as *eigenfaces* [110]. Here, PCA is used to reduce the $s \times p$ spectral matrix $D$ (with $s < p$) to a smaller $s \times s$ matrix $U$ with no loss of information. If necessary, $U$ can be further reduced with minimum loss of information.

Principal component analysis (PCA) [111] is a mathematical procedure that converts a set of observed correlated variables $\boldsymbol{x} \equiv (x_1 \ x_2 \ \ldots \ x_p)$ into a new set of uncorrelated variables $\boldsymbol{y} \equiv (y_1 \ y_2 \ \ldots \ y_p)$ via an orthogonal transformation $\boldsymbol{V}$ ($p \times p$):

$$\boldsymbol{y} = \boldsymbol{x}\boldsymbol{V}^\mathrm{T} \quad \Leftrightarrow \quad \boldsymbol{x} = \boldsymbol{y}\boldsymbol{V}, \tag{3.6}$$

with $\boldsymbol{V}\boldsymbol{V}^\mathrm{T} = \mathfrak{I}_p$ ($\mathfrak{I}$ is the $p \times p$ identity matrix). The new variables $\boldsymbol{y}$ are called *scores*, the new axes $\boldsymbol{V}$ *principal components*. The principal components are determined by a *training set* $\boldsymbol{D}$ ($s \times p$). Similarly to Eq. (3.6), the scores $\boldsymbol{U}$ ($s \times p$) of $\boldsymbol{D}$ can be computed with

$$\boldsymbol{U} = \boldsymbol{D}\boldsymbol{V}^\mathrm{T} \quad \Leftrightarrow \quad \boldsymbol{D} = \boldsymbol{U}\boldsymbol{V}. \tag{3.7}$$

The first axis (first row of $\boldsymbol{V}$) is chosen in such a way that the corresponding score (first column of $\boldsymbol{U}$) has the largest possible variance. The second axis is chosen in the same way, with the constraint that it should be orthogonal to the first axis, and so on. By construction, the rows of $\boldsymbol{V}$ form an orthogonal base. Since $s < p$, the rows of $\boldsymbol{D}$ span at most an $s$-dimensional space. Consequently, only the first $s$ columns of $\boldsymbol{U}$ are different from zero. If $\boldsymbol{x}$ is an exact linear combination of spectra from $\boldsymbol{D}$, then only the first $s$ columns of $\boldsymbol{y}$ are different from zero[2]. The last $p - s$ columns of both $\boldsymbol{y}$ and $\boldsymbol{U}$ can be discarded, as well as the last $p - s$ rows of $\boldsymbol{V}$. We can now rewrite Eq. (3.5) as

$$\boldsymbol{c} = \boldsymbol{x}\boldsymbol{D}^\mathrm{T}(\boldsymbol{D}\boldsymbol{D}^\mathrm{T})^{-1} = \boldsymbol{x}\boldsymbol{V}^\mathrm{T}\boldsymbol{U}^\mathrm{T}(\boldsymbol{U}\boldsymbol{V}\boldsymbol{V}^\mathrm{T}\boldsymbol{U}^\mathrm{T})^{-1} = \boldsymbol{y}\boldsymbol{U}^\mathrm{T}(\boldsymbol{U}\boldsymbol{U}^\mathrm{T})^{-1}. \tag{3.8}$$

Notice the similarity between Eq. (3.5) and (3.8). The advantage of using PCA to compute $\boldsymbol{U}$ becomes apparent if $\boldsymbol{U}\boldsymbol{U}^\mathrm{T}$ is ill-conditioned and cannot be inverted accurately. By construction, the first column of $\boldsymbol{U}$ has the largest variance, the second column has the second largest variance, and so on, and the last column of $\boldsymbol{U}$ has the smallest variance. The last column of

---
[2]If one of the columns $s+1,\ldots,p$ of $\boldsymbol{y}$ is significantly different from zero, then $\boldsymbol{y}$ points out of the $s$-dimensional space spanned by $\boldsymbol{D}$, i.e., $\boldsymbol{x}$ is not a linear combination (mixture) of spectra *exclusively* from $\boldsymbol{D}$.

$U$ can be iteratively removed with minimum loss of information until $UU^T$ is no longer ill-conditioned. If $DD^T$ is not ill-conditioned, then Eq. (3.5) and (3.8) will provide the same result (but the computation with Eq. (3.5) is faster because there is no PCA step).

The principal components $V$ can be computed either by eigenvalue decomposition of the covariance matrix of $D$, or by singular value decomposition of the data matrix $D$ directly. However, there still remains the problem of the negative concentrations produced by Eq. (3.5) or (3.8).

## 3.4 Improved Mix-Match Algorithm

In the previous section it was shown that given a measured spectrum $x$, the concentrations $c$ of pure substances with spectra $D$ can be computed directly [Eq. 3.5] or by using the scores $y$ of $x$ and $U$ of $D$, respectively, obtained via PCA [Eq. 3.8]. The second method has the advantage that since the scores are sorted in order of decreasing importance, one or more columns of $y$ and $U$ can be removed (starting from the last column) with minimum loss of information. Negative concentrations resulting from Eq. (3.5) or (3.8) are mathematical artifacts caused by the requirement that the norm of the residual $\epsilon = x - cD$ should be minimal (least-square principle).

The following algorithm ([112]) is based on the work of Nyden [113] and Lo and Brown [114]. The goal is to identify and quantify the components of a gas mixture based on its absorbance spectrum by comparing known spectra of pure substances to the measured spectrum. The workflow is illustrated in Fig. 3.1. The spectral library is prepared by inserting the spectra of $s$ pure substances sampled at $p$ wavelengths $\lambda_1, \lambda_2, \ldots, \lambda_p$ into the $s \times p$ matrix $D$. The wavelength points must be the same for all spectra in $D$ and $x$. If that is not the case it is necessary to compute the missing values via interpolation. The PCA is then performed on the matrix $D$, yielding the principal components $V$. Next, the scores $y$ of $x$ and $U$ of $D$ are computed with Eqs. (3.6) and (3.7), and the last $p - s$ columns of $y$ and $U$ are removed. Then, the concentrations $c$ are computed with Eq. (3.8). If $UU^T$ is ill-conditioned – this can happen if two spectra in $D$ are very similar – the last column of $U$ can be removed. Based on the computed concentrations $c = (c_1\, c_2 \ldots c_s)$, a *rating* $r_k$ ($k = 1, \ldots, s$) is assigned to each substance, defined as the maximum absorbance of the substance at the given concentration:

$$r_k \equiv c_k \max(\boldsymbol{D}_k) = c_k \max(D_{k1}\, D_{k2} \ldots D_{kp}) \qquad k = 1, \ldots, s, \qquad (3.9)$$

where $\boldsymbol{D}_k$ is the $k$-th row of the spectral library $D$. Negative concentrations $c_k$ result in negative ratings $r_k$. If all $c_k$ are positive the result is physically meaningful, but we also accept negative concentrations from Eq. (3.8) as long as the corresponding rating is close enough to zero. The reason for this is that the spectra in $D$ contain, like $x$, noise and possibly baseline offsets. If the negative absorbance values resulting from a negative concentration do not exceed typical noise and baseline drift levels, then the concentration is acceptable. If one or more ratings are below the chosen threshold, the spectrum of the substance corresponding to the lowest rating is removed from $D$, since it is unlikely that this substance is present in the measured gas sample. PCA is performed again on the new matrix $D$ and the previous steps are repeated until all ratings are above the selected threshold. Then, the substance with the *highest* rating is stored in a *hit list* together with its concentration, and an adaptive filter is applied on $D$. The filter removes the spectrum of the substance with the highest rating from $D$ and subtracts it from $x$. Additionally, it removes from every row of $D$ the projection of the top hit spectrum onto that row:

$$D_k \to D_k - \frac{D_k \cdot s}{\|s\|^2} s, \qquad (3.10)$$

where $D_k$ is the $k$-th row of $D$ and $s$ is the top hit spectrum. PCA is recomputed with the new matrix $D$ and the previous steps are repeated. When the matrix $D$ is empty (or when the hit list has reached a predefined length) the iteration is interrupted. The hit list holds the components of the mea-

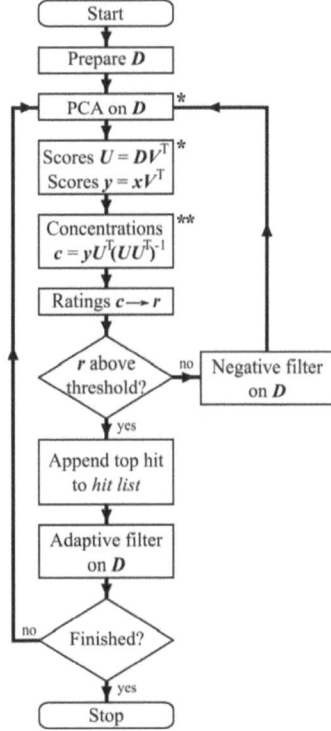

**Figure 3.1** – Workflow of the improved mix-match algorithm. If $(DD^T)^{-1}$ is not ill-conditioned, the steps indicated with a * can be skipped and $U$ replaced by $D$ in ** .

sured gas sample and their concentrations in order of decreasing rating (i.e., in order of decreasing absorbance).

As a final remark, notice that if the number of wavelength points $p$ is large compared to the number of substances $s$ in the database, then $(\mathbf{DD}^\mathrm{T})^{-1}$ in Eq. (3.5) should never be ill-conditioned. In this case it is not necessary to use PCA. The algorithm (Fig. 3.1) should still be used, as it avoids negative concentrations, but without the steps indicated with a *, and by setting $\mathbf{U} = \mathbf{D}$ in **. With PCA and with $s = 360$ spectra of approximately $p = 6000$ points each the algorithm takes about 5–10 min to complete (on a recent standard desktop computer). Without PCA it is about 10 times faster.

# Chapter 4

# Measurements on Smoke Samples Produced *In Vitro*

As a first step towards the analysis of surgical smoke collected during laparoscopic surgery, we produced smoke in the lab by cauterizing animal meat with a high-frequency electroknife in a carbon dioxide atmosphere. The electroknife is of the monopolar type similar to those used in surgery, although of an older generation. A high-frequency (> 100 kHz) voltage is applied to a small electrode tip which is brought into contact with the biological tissue. Spectra of all smoke samples were measured with our difference frequency generation (DFG) based spectrometer (see Sec. 2.4), and a few of them with a Fourier-transform infrared (FTIR) spectrometer, too [115]. The quantitative analysis was carried out with the improved mix-match algorithm (see Sec. 3.4). We investigated whether the biological tissue or the atmosphere during cauterization had any effect on the smoke composition.

## 4.1 Smoke Production and Sampling

We designed and built a simple cell that allows the cauterization of small pieces of animal meat with a high-frequency electroknife within a selected atmosphere (Fig. 4.1). The cell consists of a plexi-glass cylinder (inner $\varnothing$ 11 cm, outer $\varnothing$ 12 cm, height 5.5 cm) held between two aluminum plates (14 × 14 × 0.5 cm). Two neoprene rings are squeezed between the plexi-glass cylinder and the aluminum plates, one at each end. A 2–cm hole in the

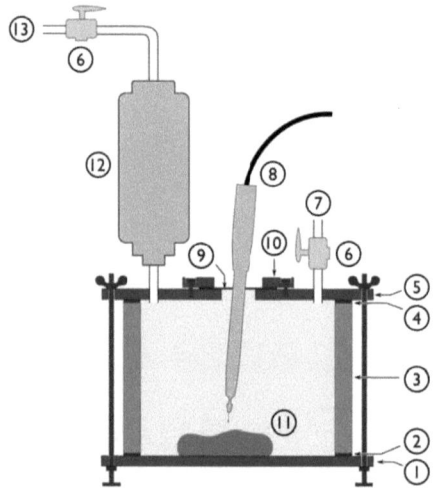

**Figure 4.1** – Schematic view of the cell used to cauterize animal meat. (1) Grounded aluminum plate; (2) neoprene ring; (3) plexi-glass cylinder; (4) neoprene ring; (5) aluminum plate; (6) valves; (7) gas inlet; (8) high-frequency electroknife, connected to generator; (9) neoprene layer with hole in the middle; (10) aluminum ring; (11) animal meat; (12) heated particle filter; (13) smoke outlet.

upper plate is covered by a thin neoprene layer (thickness 0.5 mm) with a smaller (5 mm) hole in the middle, through which the electroknife (Coagulasem, France) can be inserted. The purpose of the neoprene layer and rings is to make the cell reasonably air-tight. Two gas connections permit to flush the cell with a selected gas (e.g., carbon dioxide, nitrogen, synthetic air). At the outlet, a heated micro-glass-fiber particle filter (Infiltec GmbH, Germany, housing: model SL 215.401, filter elements: model 25-64-30) removes soot from the gas flow.

To collect and store the smoke samples, a 1 $\ell$ glass bottle was used in most cases (Fig. 4.2). The smoke samples were captured by flushing a selected gas (carbon dioxide, nitrogen or synthetic air) through the cell, particle filter and bottle while cauterizing the animal meat with the high-frequency electroknife (Fig. 4.3), and then closing the valves to seal off the bottle. To transfer the smoke sample to the evacuated high temperature multipass cell (HTMC, see Sec. 2.4.1), the bottle was placed in a water bath and connected to the HTMC via a heatable stainless steel tube (Fig. 4.4). The water in the bath was heated until it boiled and the connection tube was heated up to 150°C. The pressure inside the glass bottle was monitored

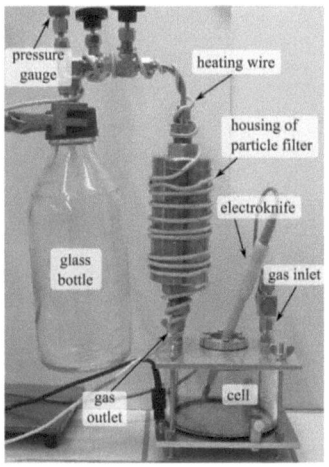

**Figure 4.2** – Smoke production cell with 1 $\ell$ glass bottle for sample collection.

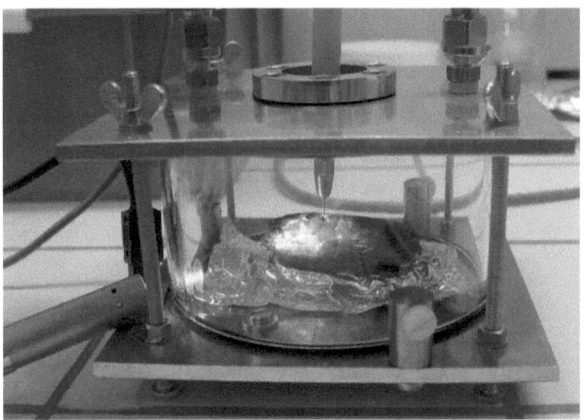

**Figure 4.3** – Animal meat being cauterized with the high-frequency electroknife.

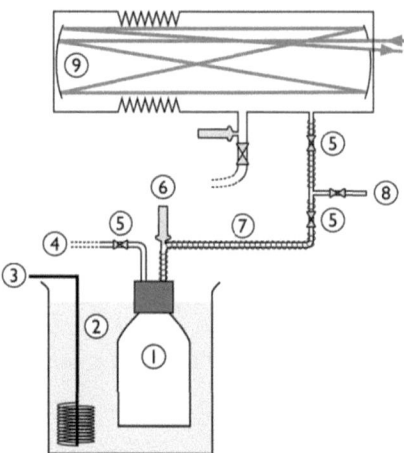

**Figure 4.4** – Procedure for filling the high-temperature multipass cell (HTMC). (1) 1 $\ell$ glass bottle; (2) water bath; (3) heater; (4) gas inlet; (5) valves; (6) pressure gauge; (7) stainless steel tube with heating wire; (8) secondary gas inlet; (9) HTMC.

with the pressure gauge, and whenever it exceeded 1.2 bar a small amount of smoke was transferred to the HTMC by opening the valves. Once the water was boiling, the valves to the HTMC were completely opened and the pressure equalized at about 350 mbar. The same gas used to produce the smoke sample was added to the HTMC, either through the glass bottle or via the secondary gas inlet, up to a final pressure of about 950–960 mbar (local atmospheric pressure). A few samples were measured at lower pressures of 200 and 300 mbar and at higher temperatures of 100 and 120°C.

## 4.2 Analysis of Smoke Samples

Table 4.1 summarizes the results obtained on all smoke samples and the conditions under which they were produced. Most samples were produced in a carbon dioxide atmosphere, to simulate the conditions in laparoscopy. Measurements were carried out mostly at ambient pressure and room temperature, although a few samples were also measured at higher temperature and lower pressure. For two samples a FTIR spectrum was measured as well.

The spectra of all the investigated samples were qualitatively similar. Figure 4.5 depicts the absorption spectrum of sample A15 measured with the DFG spectrometer at atmospheric pressure from 2900 to $3144\,\mathrm{cm}^{-1}$. Its composition was determined with the mix-match algorithm (see Sec. 3.4) and the PNNL database [109]. Hence, only any of the 360 substances contained in the database can possibly be identified. If some additional unknown component is present – as is the case here – the algorithm can generally still find the correct remaining substances, but their computed concentrations may be wrong. The concentrations of the four components water vapor ($H_2O$, 1.7%), methane ($CH_4$, 27 ppm), ethane ($C_2H_6$, 6.5 ppm) and ethylene ($C_2H_4$, 25 ppm) vary from sample to sample. Due to the large number of absorption lines of water vapor, methane and ethane (Fig. 4.5b–d), their concentrations are relatively accurate and their presence in the smoke is certain.

In Fig. 4.6 a small extract from the spectrum of another smoke sample (sample A09) is shown. All the observed absorption lines are accounted for by the four compounds water vapor, methane, ethane and ethylene. This is true for all the smoke spectra. However, in all the measured spectra there is a broad absorption between 2900 and $3000\,\mathrm{cm}^{-1}$ (Fig. 4.5a) that cannot be attributed to any of the four previously mentioned substances, nor to any other single substance contained in the PNNL database. Removing the four spectra in Fig. 4.5b–e from the measured sample spectrum yields the residual spectrum shown in Fig. 4.7a. It contains the broad absorption that can be seen in Fig. 4.5a between 2900 and 3000 cm$^{-1}$. The finer structure is due to the different resolution of the measured spectrum and the spectra that were subtracted from it. While it is possible to find a good fit to this spectrum, the number of required substances is quite large (50 in this case) but the amount of available information in the measured spectrum is limited: there aren't several narrow absorption lines with characteristic positions and widths, as was the case in Fig. 4.5 and 4.6. Instead, there is a characterless absorption without any peculiar features.

Identifying the components in Fig. 4.7a requires either a scan over a

**Table 4.1** – Summary of the measured smoke samples. The cauterized *tissue*, the atmosphere (*atm.*) during cauterization and the loss of mass due to the cauterization ($\Delta m$) are given, as well as whether a particle filter (*Filt.*) was used, and the pressure ($p$) and temperature ($T$) during the measurement.

| | | | | | | | Concentrations | | | | |
|---|---|---|---|---|---|---|---|---|---|---|---|
| ID | Tissue | Atm. | $\Delta m$ mg | Filt. | $p$ mbar | $T$ °C | $CH_4$ ppm | $C_2H_6$ ppm | $C_2H_4$ ppm | $H_2O$ % | FT[*] |
| A05 | rabbit liver | $CO_2$ | | no | 957 | 25 | 15 | 3.5 | 18 | 0.66 | |
| A06 | rabbit liver | $CO_2$ | | yes | 956 | 25 | 14 | 2.8 | 13 | 0.69 | |
| A07 | rabbit liver | s.a.[†] | 316.7 | yes | 960 | 25 | 19 | 4.0 | 20 | 0.99 | |
| A08 | beef heart | $CO_2$ | | yes | 960 | 25 | 6.4 | 0.7 | 7.5 | 0.37 | |
| A09 | veal heart | $CO_2$ | 394.7 | yes | 960 | 25 | 11 | 2.1 | 7.1 | 0.74 | |
| A10 | beef heart | $CO_2$ | 769.4 | yes | 967 | 100 | 4.2 | 0.8 | 9.3 | 1.5 | |
| A11 | pig heart | $CO_2$ | | yes | 900 | 100 | 13 | 3.0 | 19 | 1.3 | |
| A12 | beef loin | $CO_2$ | 292.6 | no | 200 | 100 | 12 | 5.5 | ≈15 | 0.35 | |
| A13 | veal heart | $CO_2$ | 269.1 | yes | 300 | 100 | 14 | 3.8 | ≤10 | 0.24 | |
| A14 | beef liver | $CO_2$ | 556.1 | yes | 300 | 120 | 17 | 6.3 | ≤10 | 1.2 | |
| A15 | beef liver | $CO_2$ | 352.8 | yes | 930 | 25 | 27 | 6.5 | 25 | 1.7 | |
| A16 | pig liver | $N_2$ | 669.0 | yes | 930 | 25 | 36 | 9.5 | 37 | 2.3 | |
| A17 | pig | $CO_2$ | | yes | 930 | 25 | 29 | 6.1 | 17 | 1.1 | ■ |
| A19 | pig kidney | $CO_2$ | 686.3 | yes | 930 | 25 | 41 | 11 | 32 | 0.27 | ■ |
| A20 | pig kidney | $CO_2$ | | yes | 930 | 25 | 34 | 6.5 | 11 | 0.49 | |
| Average | | | | | | 20 | 4.8 | 17 | 0.87 | | |
| Min.–Max. | | | | | | | 4.2–41 | 0.7–11 | 7.1–37 | 0.15–23 | |

[*] A Fourier-transform infrared (FTIR) spectrum was measured for entries marked with ■.
[†] Synthetic air.

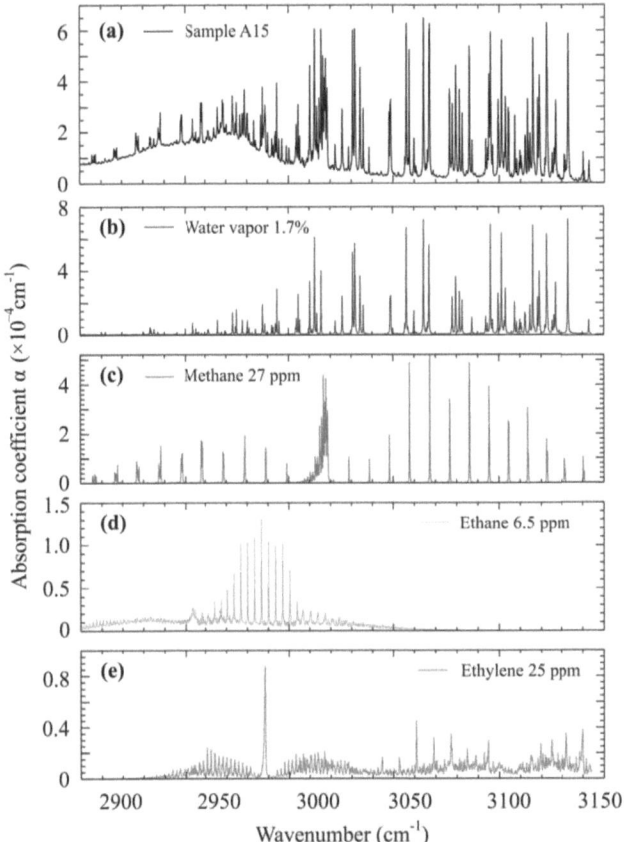

**Figure 4.5** – **(a)** Absorption spectrum of smoke sample A15 measured with the DFG spectrometer. **(b)–(e)** Spectra of the four identified components water vapor (1.7%), methane (27 ppm), ethane (6.5 ppm) and ethylene (25 ppm) taken from the PNNL database [109].

larger wavelength range, or a measurement at lower pressure in order to reduce pressure broadening and make single absorption lines visible. The second approach works, for example, for benzene, which has a very broad absorption band at ambient pressure (Fig. 4.8) that can be resolved into individual absorption lines at low pressure ($p \approx 15$ mbar) [116]. One must consider, however, that the resolution is limited by the linewidth of the idler (150 MHz, Tab. 2.1), and that the wavenumber step size of the spectrometer lies between $0.002\,\mathrm{cm}^{-1}$ (60 MHz) and $0.020\,\mathrm{cm}^{-1}$ (600 MHz). One sample (sample A12) was measured at 200 mbar and two (samples A13 and A14) at

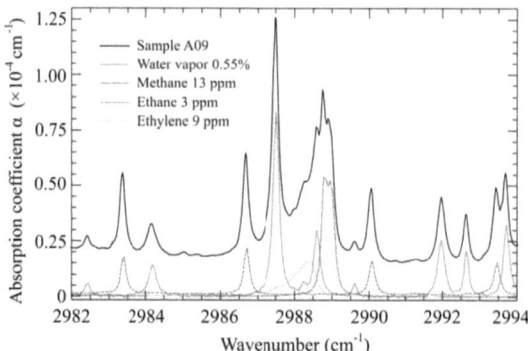

**Figure 4.6** – Extract of the spectrum of a smoke sample (solid line) and spectra of the four identified components (water, methane, ethane and ethylene) taken from the PNNL database [109].

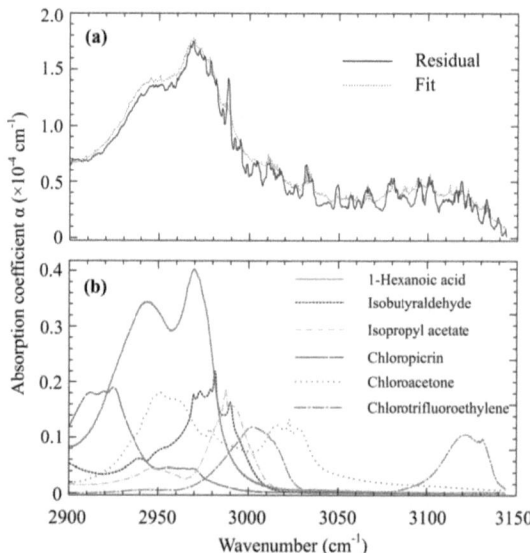

**Figure 4.7** – (a) Residual of the spectrum in Fig. 4.5a (solid line) after subtracting water vapor, methane, ethane and ethylene, and spectrum computed using the mix-match algorithm (see Sec. 3.4) with the PNNL database (dotted line). (b) Six out of the 50 substances needed for the fit in (a). Notice that the scales on the vertical axes in (a) and (b) are different.

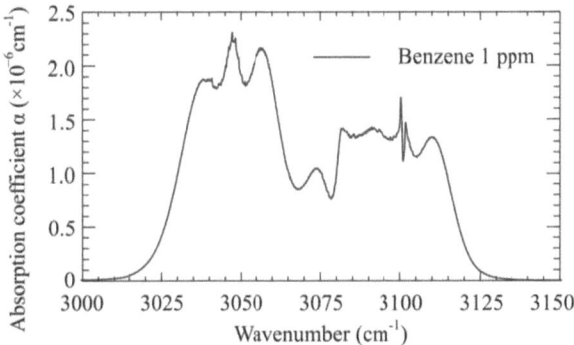

**Figure 4.8** – Absorption spectrum of benzene (1 ppm) buffered in nitrogen between 3000 and 3150 cm$^{-1}$ at ambient pressure and room temperature (from PNNL [109]).

300 mbar. At these pressures, the PNNL database can no longer be used, so instead the spectra of water vapor, methane, ethane and ethylene were computed based on data from the HITRAN database [107]. The measured spectrum was overlaid to the four spectra and visually inspected across the entire wavelength range to find further absorption lines that would point to one or more additional substances, but none were found (Fig. 4.9). Hence, we could not find any explanation for the broad absorption (Fig. 4.7) that ap-

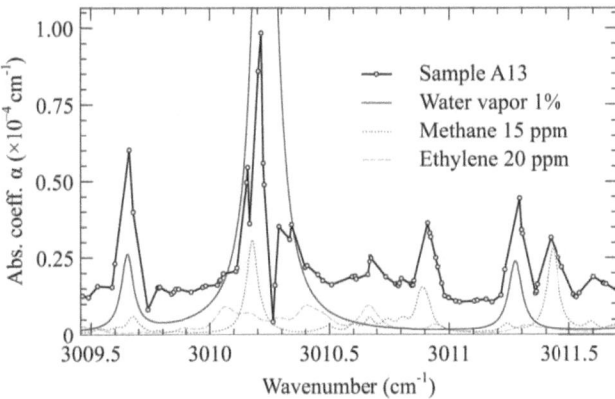

**Figure 4.9** – Absorption spectrum of a smoke sample recorded at 300 mbar and 120°C and absorption spectra of water vapor, methane and ethylene at the same temperature and pressure from HITRAN [107].

pears in all measurements. Water absorption lines are problematic as they appear everywhere (Fig. 4.5b). At low pressure, distortion of absorption line shapes – such as the water absorption line at 3010.25 cm$^{-1}$ in Fig. 4.9 – appear due to the too large wavenumber step size and to the fact that the same absorption line was also present in the reference measurement. It is possible that very narrow and weak absorption lines lying between two sampling points could be missed.

The absorption spectra of samples A17 and A19 were measured both with the DFG spectrometer and with a FTIR spectrometer (Bruker Optics Inc., U.S.A., model IFS 66v). The spectrum of sample A17 is shown in Fig. 4.10. Additional substances which do not absorb in the range accessible with the DFG spectrometer could be identified: nitric oxide NO (25 ppm), carbon monoxide CO (200 ppm), nitrous oxide $N_2O$ (50 ppm), acetylene $C_2H_2$ (45 ppm), and hydrogen cyanide HCN (30 ppm). Carbon dioxide and water absorption lines saturate the absorption in the intervals 1320–1910 cm$^{-1}$ (water), 2240–2380 cm$^{-1}$ (carbon dioxide), and 3530–3960 cm$^{-1}$ (both). The inset in Fig. 4.10a shows the spectrum of the same sample measured with the DFG spectrometer. Some water absorption lines are missing in the FTIR spectrum due to the lower sensitivity and resolution of the FTIR spectrometer (0.125 cm$^{-1}$) compared to the DFG spectrometer ($\leq 0.02$ cm$^{-1}$). Figures 4.10b–d show magnifications of the FTIR spectrum and spectra from the PNNL database of the identified compounds. The qualitative composition of sample A19 was the same, the concentrations were slightly different.

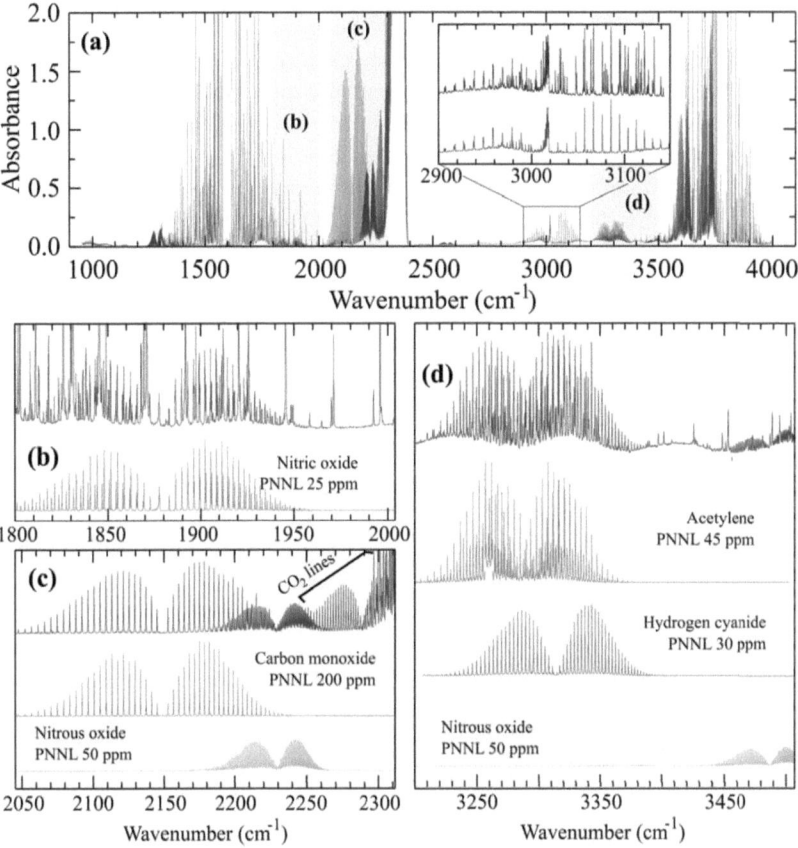

**Figure 4.10** – (a) FTIR spectrum (resolution 0.125 cm$^{-1}$, absorption pathlength 4 m) of smoke sample A17. Inset: The spectral range 2900–3150 cm$^{-1}$ measured with the DFG spectrometer compared to the FTIR spectrum. (b) Nitric oxide absorption lines between 1800 and 2000 cm$^{-1}$. (c) Carbon monoxide and nitrous oxide absorption lines between 2050 and 2300 cm$^{-1}$. (d) Acetylene, hydrogen cyanide and nitrous oxide absorption lines between 3200 and 3500 cm$^{-1}$.

## 4.3 Discussion

In all of the 15 measured smoke samples (Tab. 4.1) the same four compounds were detected: water vapor, methane, ethane and ethylene [115]. Concentrations were in the ppm range for methane, ethane and ethylene, and in the promille/percent range for water vapor. FTIR spectra taken of two of the samples additionally revealed nitric oxide (25 ppm), carbon monoxide (200 ppm), nitrous oxide (50 ppm), acetylene (45 ppm) and hydrogen cyanide (30 ppm).

The first sample (sample A05) was measured without a particle filter. When no explanation could be found for the broad absorption that appeared at 2900–3000 cm$^{-1}$ (Fig. 4.7a) we thought it could be due to scattering on soot particles. Hence, all following samples (except one, A12) were filtered with a particle filter retaining particles with diameter $\geq 0.1$ $\mu$m, but the absorption still remained, as can be seen, for example, in Fig. 4.5. Furthermore, one would not expect scattering to be significantly different at 2900 cm$^{-1}$ from 2950 cm$^{-1}$ (Fig. 4.7a). The particle filter was used nevertheless to keep the HTMC and its mirrors clean.

The concentrations of the components in the smoke sample depend on how much animal tissue was cauterized. Figure 4.11 shows the methane concentration, and the ethane/methane and ethylene/methane concentration ratios for four pairs of smoke samples. The methane concentration appears to be slightly higher for the two smoke samples produced by cauterizing pig kidney (samples A19, A20). There is no obvious correlation between the ethane/methane and ethylene/methane concentration ratios in smoke samples produced from different tissues.

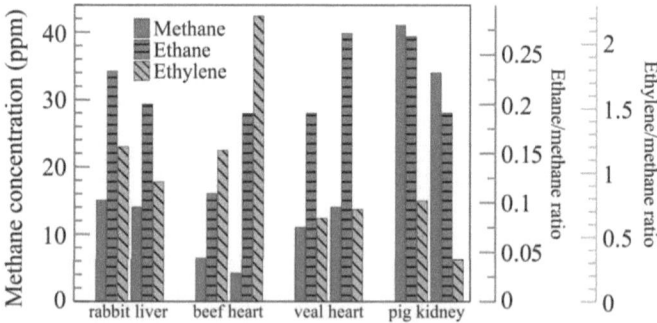

**Figure 4.11** – Methane concentration, and ethane and ethylene concentration relative to methane for four pairs of smoke samples produced by cauterizing different tissues.

Sample A07 and A16 were produced in a synthetic air and nitrogen atmosphere, respectively, and all others in carbon dioxide. The compositions of samples A07 and A16 were not significantly different from the other samples (Fig. 4.12).

Detection thresholds can be computed by using the smallest measurable change in absorption coefficient given in Tab. 2.1 on page 60 or, for substances that have resolvable absorption lines at atmospheric pressure, with the noise equivalent absorption coefficient given in the same table. Either method, however, does not take into account interference with other substances present in the sample. For example, 1 ppm of benzene has a peak absorption coefficient of $2.3 \times 10^{-6}$ cm$^{-1}$ near 3050 cm$^{-1}$ (Fig. 4.8), well above the minimum measurable absorption coefficient of $8.7 \times 10^{-7}$ cm$^{-1}$. However, if its spectrum is added to the spectrum of a smoke sample, it is clear that such a small concentration is not measurable, as shown in Fig. 4.13. Based on Fig. 4.13, we choose the minimum measurable absorption coefficient as $\alpha_{min} = 2 \times 10^{-5}$ cm$^{-1}$. Notice that this is a factor 23–40 times larger than in the interference-free case (Tab. 2.1). The computed detection thresholds for several compounds are listed in Tab. 4.2. The data in Tab. 4.2 are conservative estimates. Compounds with narrow absorption lines – such as methane, ethane, ethylene, hydrogen chloride, formaldehyde – have lower detection thresholds. A systematic approach to determine detection thresholds would be to repeatedly add a small amount of an undetected substance to a smoke sample and determine at which concen-

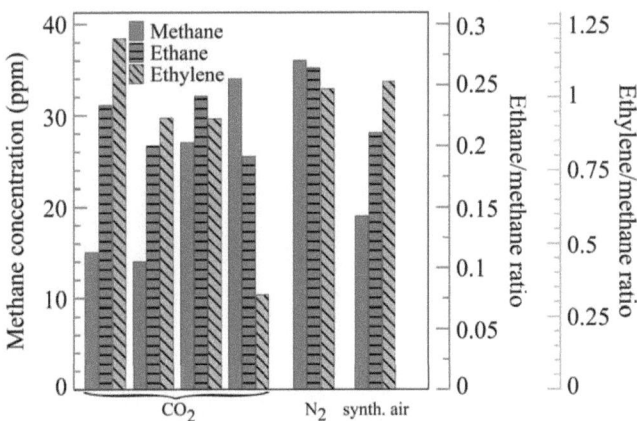

**Figure 4.12** – Methane concentration, and ethane and ethylene concentration relative to methane for six smoke samples produced in carbon dioxide, nitrogen and synthetic air.

**Table 4.2** – A few selected compounds with their detection thresholds with the DFG spectrometer (assuming the "background" spectrum given in Fig. 4.13) and recommended exposure limits (REL) [16]. *Emphasized* entries are detectable with the DFG spectrometer at concentrations below the REL.

| Substance | $c_{min}$ ppm | REL ppm | Substance | $c_{min}$ ppm | REL ppm | Substance | $c_{min}$ ppm | REL ppm |
|---|---|---|---|---|---|---|---|---|
| **(Cyclo) alkanes, -enes, -ynes** | | | **Nitriles** | | | **Inorganics** | | |
| *methane* | 1.0 | 10000 | acetonitrile | 32 | 20 | *hydrogen chloride* | 0.9 | 2 |
| *ethane* | 1.0 | 10000 | acrylonitrile | 150 | 2 | nitrogen dioxide | 11 | 3 |
| *propane* | 1.6 | 1000 | | | | water | 450 | — |
| *butane* | 2.2 | 800 | **(Cyclo) alcohols** | | | hydrogen fluoride | — | 1 |
| *pentane* | 2.1 | 600 | methanol | 7.8 | 200 | | | |
| *hexane* | 2.0 | 50 | ethanol | 6.9 | 500 | **Aromatic compounds** | | |
| *octane* | 1.4 | 300 | propanol | 4.5 | 200 | benzene | 8.7 | 0.5 |
| *nonane* | 1.1 | 200 | isopropanol | 3.0 | 200 | *m-xylene* | 7.7 | 100 |
| decane | 0.9 | — | cyclohexanol | 1.2 | 50 | *o-xylene* | 7.6 | 100 |
| undecane | 0.8 | — | | | | *p-xylene* | 6.5 | 100 |
| *ethylene* | 5.0 | 10000 | **Aldehydes** | | | styrene | 11 | 20 |
| *1,3-butadiene* | 18 | 5 | formaldehyde | 0.78 | 0.3 | *toluene* | 8.8 | 50 |
| hexene | 3.8 | — | acetylaldehyde | 50 | 50 | furan | 130 | — |
| *cyclohexane* | 0.5 | 200 | benzaldehyde | 15 | — | *pyridine* | 12 | 5 |
| *cyclohexene* | 1.4 | 300 | acrolein | 65 | 0.1 | **Others** | | |
| | | | | | | *acetone* | 14 | 500 |
| **Amines** | | | **Carboxylic acids** | | | *dichloromethane* | 67 | 50 |
| methylamine | 10 | 10 | formic acid | 7.7 | 5 | nicotine | 2.7 | 0.07 |
| dimethylamine | 6.4 | 2 | acetic acid | 32 | 10 | sevoflurane | 20 | 2* |
| trimethylamine | 3.8 | 2 | | | | ammonia | 750 | 20 |
| | | | | | | carbon monoxide | — | 30 |

*Generally recommended exposure limit for all halogenated anesthetics.

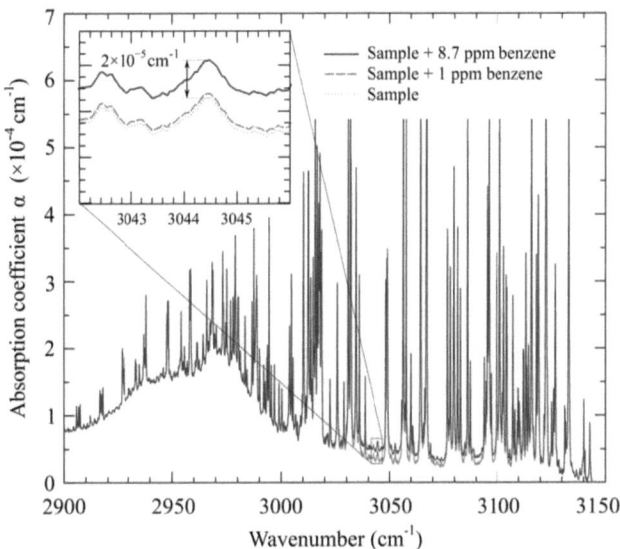

**Figure 4.13** – Spectrum of sample A15 plus 1 ppm benzene and plus 8.7 ppm benzene overlaid to the spectrum of sample A15. Inset: magnification around $3044\,\text{cm}^{-1}$ where benzene absorbs strongly.

tration said substance becomes detectable. This, however, would have to be repeated for every single substance several times, a very tiresome experiment.

Most alkanes/alkenes, alcohols and some aromatic compounds can be detected at concentrations well below the recommended exposure limit (REL). Table 4.2 does not include substances that do not absorb significantly between 2900 and $3144\,\text{cm}^{-1}$. To this group belong, among others, the five compounds detected with FTIR spectroscopy: hydrogen cyanide (30 ppm, REL 1.9 ppm), nitric oxide (25 ppm, REL 25 ppm), nitrous oxide (50 ppm, REL 100 ppm), carbon monoxide (200 ppm, REL 30 ppm) and acetylene (45 ppm, REL 1000 ppm). It is interesting that none of the compounds found frequently in previous studies – toluene, benzene, xylene, ethylbenzene, Tab. 1.2 – could be detected here.

# Chapter 5

# Measurements on Surgical Smoke

## 5.1 Sample Collection

In laparoscopic surgery (minimally invasive, keyhole surgery [117]), the abdominal cavity of the patient is filled with a suitable gas (usually carbon dioxide [19]). Several trocars are inserted through the abdominal wall and provide access points for the laparoscope and other surgical tools (Fig. 5.1). Bipolar vessel sealing devices are often used to cut through tissue with minimum bleeding [18]. An insufflator maintains the pressure inside the abdominal cavity at 15–20 mbar above ambient pressure. The surgeon can periodically open the valve on the trocar to evacuate smoke and improve visibility. A gas sample bag can thus be filled without requiring a pump by connecting it to one of the trocars. For this study we used transparent Tedlar bags (CEL Scientific, U.S.A.) with $3\,\ell$ capacity and a polypropylene fitting. This material has been shown to be well suited for storing gas samples, as there are only moderate losses and, except for two substances (phenol and N,N-dimethylacetamide), there is no contamination by residues left over from the manufacturing process [118].

After the smoke samples were collected during routine colorectal surgery at the University hospital Zurich (USZ), they were brought to our lab and filled into the high temperature multipass cell (HTMC, Sec. 2.4) for measurements with the DFG spectrometer (Sec. 5.2.1), or in the multipass cell (MPC, Sec. 2.5) for measurements with the DFB spectrometer (Sec. 5.2.3).

The filling procedure was the same for both cells. The cell was rinsed with nitrogen (purity 5.0) and then evacuated. The sample bag was con-

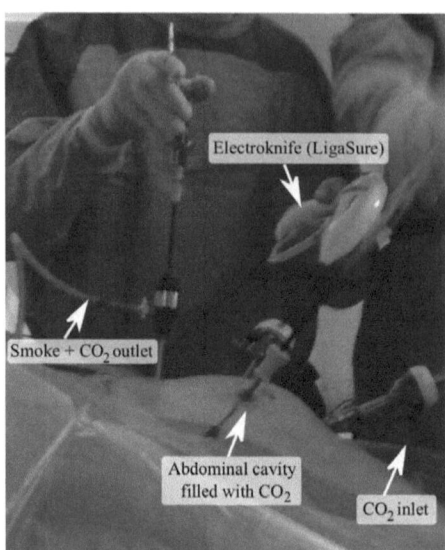

**Figure 5.1** – Picture taken during laparoscopic surgery at the University hospital Zurich (USZ). The sample bag (not shown) is connected to the tube labeled *Smoke+CO$_2$ outlet*.

nected via a Teflon tube to the evacuated cell, which could then be filled by opening a valve and letting the atmospheric pressure compress the sample bag. The duration between the sample collection and measurement ranged from two hours to a few days.

## 5.2 Results and Discussion

An overview of the measured samples is given in Tab. 5.1. A total of 34 sample bags were filled with smoke from six different operations (H01–H06). The contents of most of them were analyzed with the DFG spectrometer, similarly to the smoke samples produced in the lab (see Ch. 4) [119]. Samples from the last two operations (H05–H06) were also measured with the DFB spectrometer to measure hydrogen fluoride and carbon monoxide concentrations. One smoke sample was also measured with a FTIR spectrometer. The spectra of the samples that were measured with the DFG spectrometer were evaluated with the improved mix-match algorithm (see Sec. 3.4), while carbon monoxide and hydrogen fluoride concentrations in the samples measured with the DFB spectrometer were computed with the calibration outlined in Sec. 2.5.3.

**Table 5.1** – Overview of the samples collected at the University hospital Zurich. For each operation (*Op.*) H01–H06 several sample bags were filled (a,b,c...). The spectrometer they were measured with is indicated with a ■ in the respective column.

|     |     |         | Measured with |      |     |                           |
| --- | --- | ------- | --- | ---- | --- | ------------------------- |
| Op. | Bag | Content | DFG | FTIR | DFB | Remarks                   |
| H01 | a   | $CO_2$  | ■   | ■    |     | medical $CO_2$            |
|     | b–e | smoke   | ■   |      |     |                           |
|     | f   | smoke   | ■   | ■    |     |                           |
| H02 | a–f | smoke   | ■   |      |     |                           |
| H03 | a   | $CO_2$  | ■   |      |     | medical $CO_2$ flushed through abdominal cavity of patient |
|     | b–f | smoke   | ■   |      |     |                           |
| H04 | a–i | smoke   | ■   |      |     |                           |
|     | TD  | smoke   | ■   |      |     | preconcentrated with thermal desorption |
| H05 | a–b | smoke   |     |      | ■   |                           |
|     | c   | smoke   | ■   |      | ■   |                           |
| H06 | a   | smoke   | ■   |      | ■   |                           |
|     | b–c | smoke   |     |      | ■   |                           |

## 5.2.1 Measurements with the DFG Spectrometer

To assess the purity of the medical $CO_2$ gas bottle used at the hospital and to confirm that the employed Tedlar bags do not contaminate the samples, 3 $\ell$ of medical $CO_2$ (sample H01a) were taken from the outlet of the insufflator at the hospital and measured. The spectrum is depicted in Fig. 5.2. A close inspection reveals that only water vapor absorption lines can be detected in this spectrum. A FTIR spectrum only showed carbon dioxide in addition to water vapor. Moisture is added to $CO_2$ because dry $CO_2$ is known to potentially cause hypothermia [120].

In a later operation, 3 $\ell$ of $CO_2$ were flushed through the patient's abdominal cavity and into the Tedlar bag (sample H03a, Fig. 5.3). No electrosurgical devices had been used up to that point. The spectrum of the sample was recorded over the full range of the spectrometer, from 2817 to 2920 cm$^{-1}$ using the $\Lambda$ = 29.5 $\mu$m poling period of the PPLN crystal, and from 2900 to 3144 cm$^{-1}$ using the $\Lambda$ = 29.9 $\mu$m poling period. All the detected absorption lines are due to water vapor and no additional compound could be detected.

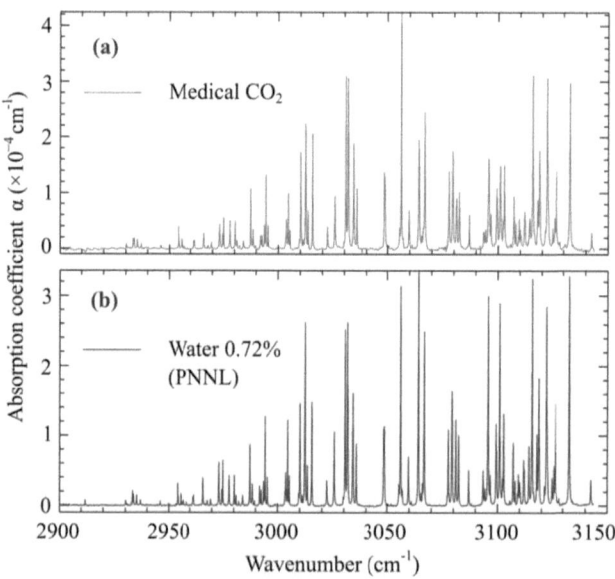

**Figure 5.2** – **(a)** Absorption spectrum of medical $CO_2$ sampled from the gas insufflator at the hospital measured at $p$ = 955 mbar and $T$ = 25 °C. **(b)** Absorption spectrum of water vapor (at 0.72%) from the PNNL database [109].

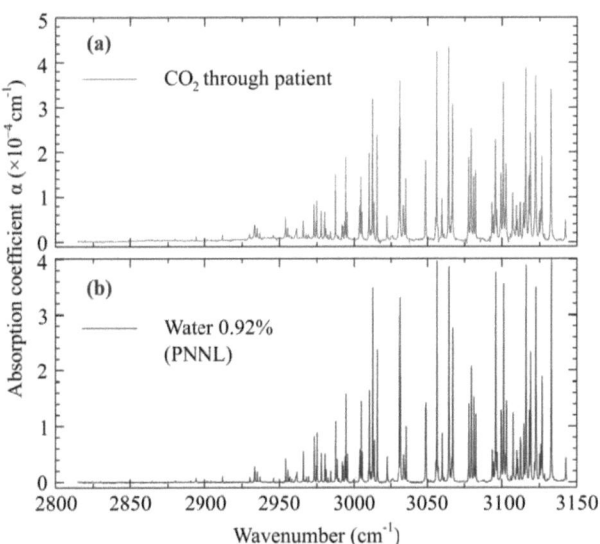

**Figure 5.3** – (a) Absorption spectrum of carbon dioxide flushed through the patient's abdominal cavity measured at $p = 960$ mbar and $T = 25$ °C. (b) Absorption spectrum of water vapor (at 0.92%) from the PNNL database.

The spectrum of a surgical smoke sample (sample H03f) is shown in Fig. 5.4. It was measured over the full tuning range of the spectrometer (2817–3144 cm$^{-1}$). Only water vapor (0.78%, Fig. 5.4b) and methane (450 ppb, not shown) could be detected. The discrepancies in the measured (Fig. 5.4a) and water vapor (Fig. 5.4b) spectrum produce artifacts in the residual spectrum (Fig. 5.4c), due to the narrow linewidth of the absorption lines and to the finite tuning step size of the spectrometer. This sample is qualitatively similar to sample H03a (Fig. 5.3a), where no surgical smoke had been produced.

The spectrum of a sample taken from a different operation (sample H02f) is shown in Fig. 5.5. This sample is very similar in composition to the smoke samples produced in the lab (see Ch. 4), with methane (9.1 ppm), ethane (2.0 ppm) and ethylene (10 ppm) (Fig. 5.5b). Additionally, there are four broad absorption features that were not observed before, which we determined to be caused by sevoflurane. Sevoflurane ($C_4H_3F_7O$, CAS Nr. 28523-86-6) is a halogenated ether used as an anaesthetic [121]. It is liquid at room temperature but very volatile (boiling point 58.6 °C, vapor pressure 263 mbar at 25 °C). As its spectrum is not in the PNNL database [109] and we couldn't find it anywhere else, we obtained a bottle from the manufac-

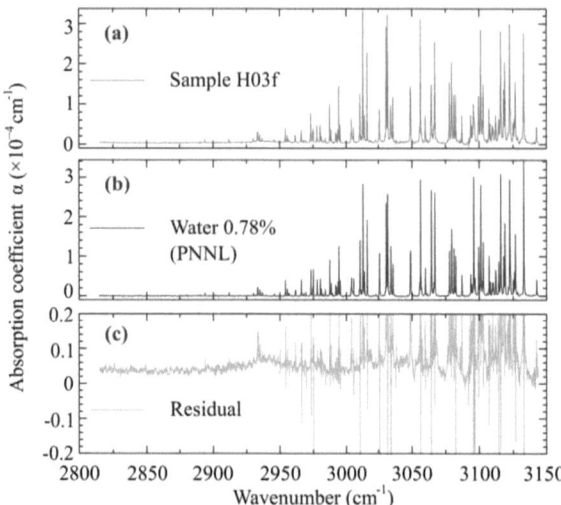

**Figure 5.4** – (a) Absorption spectrum of surgical smoke sample H03f measured at $p = 950$ mbar and $T = 25\,°C$. (b) Absorption spectrum of water vapor (at 0.78%) from the PNNL database. (c) Residual spectrum after subtraction of the water vapor spectrum (b) from (a).

turer (Abbott, U.S.A., Sevorane) and measured the spectrum of sevoflurane vapor with the DFG spectrometer separately.

The sample was prepared by injecting a small volume $V_s$ of liquid sevoflurane with a microliter syringe into the device depicted schematically in Fig. 5.6, while flushing nitrogen (purity 5.0) through it and into the previously evacuated HTMC. The volume of liquid sevoflurane $V_s$ needed to reach a final concentration $c$ in the HTMC at the temperature $T$ and total pressure $p$ can be computed from

$$c = \frac{n_s}{n_0} = \frac{\rho_s V_s R T}{p V m_s^{\text{mol}}}, \tag{5.1}$$

where $n_s, n_0$ are the number of sevoflurane and "air" moles, respectively, $\rho_s$ is the density of liquid sevoflurane and $m_s^{\text{mol}}$ is the molar mass of sevoflurane. In Eq. (5.1) it was assumed that air is an ideal gas ($pV = n_0 RT$). With the volume of the HTMC ($V = 2\,\ell$), the density of (liquid) sevoflurane $\rho_s = 1.52\,\text{g/cm}^3$ and the molar mass of sevoflurane $m_s^{\text{mol}} = 200.055\,\text{g/mol}$ we can rewrite Eq. (5.1) as

$$V_s\,(\mu\ell) = 3.17 \cdot \frac{c\,(\text{ppm}) \cdot p\,(\text{bar})}{T\,(\text{K})}. \tag{5.2}$$

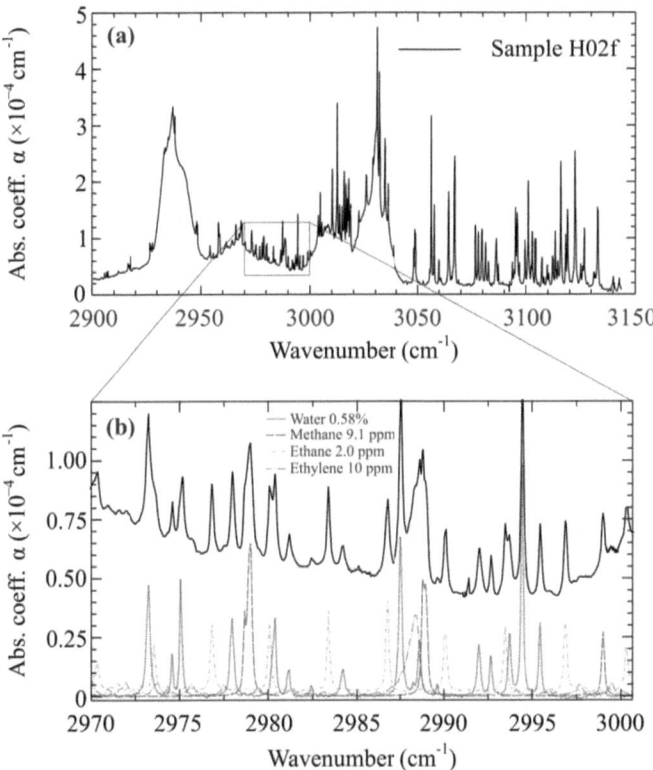

**Figure 5.5** – (a) Absorption spectrum of surgical smoke sample H02f measured at $p = 950$ mbar and $T = 25$ °C. (b) Absorption spectra of water vapor, methane, ethane and ethylene from the PNNL database.

For example, for a final concentration of 1000 ppm at $p = 0.95$ bar and at room temperature ($T = 298$ K) 10.1 $\mu\ell$ of liquid sevoflurane are needed.

The sevoflurane vapor spectra for concentrations of 1000 ppm, 500 ppm, 250 ppm and 125 ppm are shown in Fig. 5.7. The narrow absorption lines are due to water vapor. The absorption is saturated for the first (1000 ppm) measurement. With a total absorption pathlength of 35 m an absorption coefficient $\alpha > 8.6 \times 10^{-4}$ cm$^{-1}$ corresponds to a transmittance $\mathcal{T} < 0.1\%$. We manually removed the water absorption lines and then added the sevoflurane spectrum to the PNNL database so it could be used by the mix-match algorithm. The measured sevoflurane concentrations vary from $< 20$ ppm to 450 ppm. The recommended exposure limit for all halogenated anesthetics is 2 ppm [122]. During general anesthesia, sevoflurane is administered via

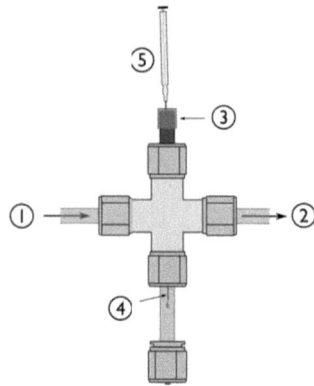

**Figure 5.6** – Device used for the preparation of sevoflurane vapor samples. (1) Gas inlet; (2) gas + vapor outlet (to the HTMC); (3) septum; (4) injection needle; (5) microliter syringe.

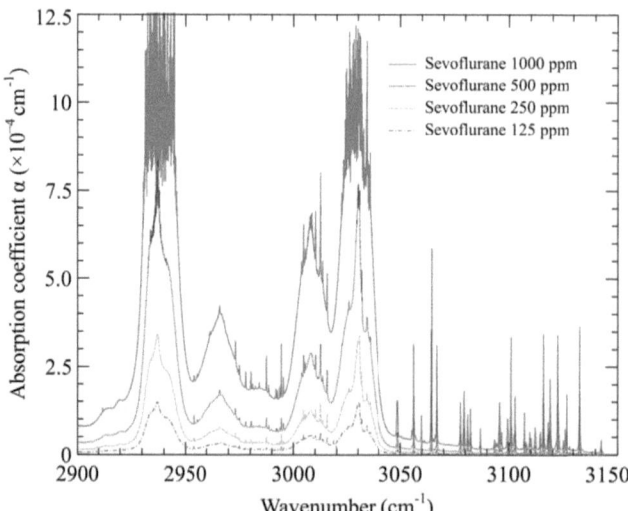

**Figure 5.7** – Absorption spectrum of sevoflurane at four different concentrations (1000 ppm, 500 ppm, 250 ppm and 125 ppm) measured at $p = 950$ mbar and $T = 25$ °C. The narrow absorption lines are due to water vapor. Absorptions larger than $8.7 \times 10^{-4}$ cm$^{-1}$ correspond to a transmittance $\mathcal{T} < 0.1\%$ and cannot be measured accurately.

inhalation at concentrations in the percent range [121]. It is not clear by which mechanism it reaches the abdominal cavity of the patient. It should also be noted, that surgeons occasionally open a valve on one of the trocars to vent smoke and improve visibility. By doing so, sevoflurane vapor is released into the operation room.

We measured one surgical smoke sample once by filling it directly into the HTMC, and once by previously filtering it with a particle filter (Infiltec GmbH, Germany, model DIF-BN 30 K), which removes particle with diameter larger than 0.1 µm from the gas flow. As can be seen in Fig. 5.8, which shows the spectra of the filtered and unfiltered sample, there is no significant difference between the two spectra.

Two different samples but originating from the same operation (sample H01f and H01e) were measured four weeks apart, and only a small difference between the two was found (Fig. 5.9). In particular, methane was not detected in the first measurement (sample H01f), but was detected at 1.1 ppm in the second one (sample H01e). This corresponds approximately to the atmospheric concentration, and could therefore be due to methane leaking into the sample bag. It could equally be due to methane being present in the sample H03e in the first place. The sevoflurane concentration is slightly lower in the second measurement (120 ppm versus 150 ppm

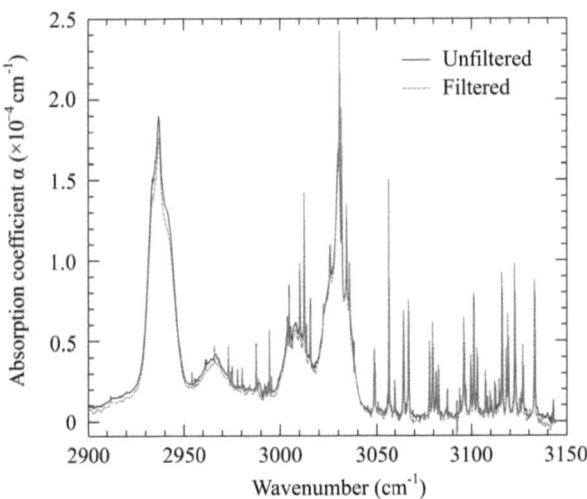

**Figure 5.8** – Absorption spectrum of a surgical smoke sample without (solid) and with (dashed) previously filtering with a particle filter (particles with $\varnothing >$ 0.1 µm are retained in the filter) measured at $p = 950$ mbar and $T = 25$ °C.

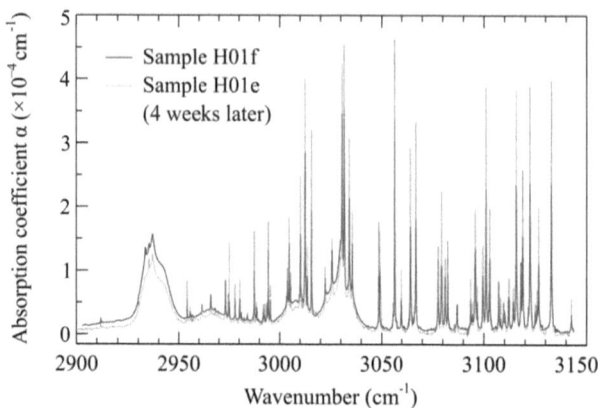

**Figure 5.9** – Absorption spectra of two samples measured four weeks apart measured at $p = 945$ mbar and $T = 25$ °C. Sample H01f (solid) was measured first, sample H01e (dotted) was measured four weeks later.

in the first one). It should be noticed, however, that since the spectra belong to two different samples, we do not expect the original concentrations to be equal. However, it is clear that there is no dramatic loss or change in the gas mixture even over a period of a month.

## Measurements with Thermal Desorption

Given the multitude of chemical species found in previous studies (Tab. 1.2), it is reasonable to assume that our sensitivity is insufficient. In an effort to increase the sensitivity and as a possible alternative sampling technique we explored the possibility of using thermal desorption tubes. A thermal desorption tube is filled with an adsorber to which certain compounds stick (adsorb) very easily at room temperature, but are promptly desorbed at higher temperature. Thermal desorption (TD) tubes act as filters by retaining certain substances present in a gas flow, but not the buffer gas itself. They can be used to collect samples by establishing a slow continuous flow through them. Later, the retained substances can be desorbed by heating the adsorber and flushing the tube with a pure gas. The advantage lies in the fact that the gas volume used for the desorption can be chosen to be orders of magnitude smaller than the total gas volume used during the sample collection (adsorption). This dramatically increases the concentrations and is useful, for example, in analytical techniques that only work with very small sample volumes, such as gas chromatography. More generally, it is useful whenever the sample volume exceeds the maximum measurable volume, because the contents of a large volume sample can be adsorbed and then desorbed into a smaller (measurable) volume at higher concentrations. Thermal desorption tubes can also be used to collect samples in the first place. The main advantage consists in the smaller volume of the "sample" (compared to using sample bags). The drawback is that adsorbers are selective: while a sample bag holds every substance present in the gas mixture, an adsorber will only be effective with certain groups of chemicals. Thermal desorption tubes aimed at storing a wide range of compounds usually contain several *beds*, with each one offering good retention properties for a different subset of chemical compounds.

We built a thermal desorption tube, shown in Fig. 5.10, made of a stainless steel tube ($\varnothing$ 10 mm, length 12 cm) and filled with 2.877 g Carboxen 569 (Supelco, U.S.A.), a carbon molecular sieve. At both ends two silanized glass wool plugs keep the Carboxen spheres (mesh 20/45) in place. The tube is heatable up to nearly 400 °C thanks to the heating wire (Horst GmbH, Germany, model HS1) wound around the steel tube along its entire length. The temperature can be monitored by means of a thermistor clamped between the steel tube and the heating wire. The breakthrough volumes[1] of Carboxen 569 for a few selected compounds are given in Tab. 5.2. The emphasized entries denote substances with breakthrough volumes larger than 25 $\ell$ at 20 °C and smaller than 2 $\ell$ at 360 °C. For these we expect an increase

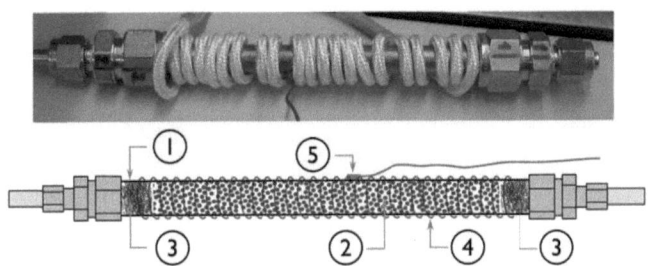

**Figure 5.10** – Home-built thermal desorption tube. (1) Stainless steel tube (length 12 cm, inner $\varnothing$ 8 mm, outer $\varnothing$ 10 mm); (2) 2.877 g Carboxen 569; (3) silanized glass wool plugs; (4) heating wire (length 1 m, 350 W); (5) thermistor.

---

[1]The *breakthrough volume* is the carrier gas volume required to elute a specified compound off 1 g of adsorber. For example, acetone has a breakthrough volume of 7 $\ell$ at 20 °C. This means that if a small amount of acetone is injected into the desorption tube and a carrier gas flow is switched on, then acetone will start to exit the desorption tube after 7 $\ell$ of gas have passed through. An adsorber is suitable for a compound if the breakthrough volume is large at low temperature (good retention), and small at high temperature (good recovery).

**Table 5.2** – Breakthrough volumes for 1 g of Carboxen 569 for selected compounds at 20, 200 and 360 °C. *Emphasized* entries denote compounds with breakthrough volumes larger than 25 $\ell$ at 20 °C and smaller than 2 $\ell$ at 360 °C. Source: Scientific Instrument Services.

| Substance | Breakthrough volume at | | | Substance | Breakthrough volume at | | |
|---|---|---|---|---|---|---|---|
| | 20 °C | 200 °C | 360 °C | | 20 °C | 200 °C | 360 °C |
| Methane | 0.035 $\ell$ | 0.003 $\ell$ | — | Ethanol | 3.3 $\ell$ | 0.070 $\ell$ | 0.008 $\ell$ |
| Ethane | 0.50 $\ell$ | 0.011 $\ell$ | — | Acetone | 7.0 $\ell$ | 0.040 $\ell$ | 0.002 $\ell$ |
| Methanol | 0.95 $\ell$ | 0.035 $\ell$ | 0.005 $\ell$ | Propanol | 12 $\ell$ | 0.16 $\ell$ | 0.010 $\ell$ |
| Propane | 3.0 $\ell$ | 0.024 $\ell$ | 0.001 $\ell$ | Isopropanol | 16 $\ell$ | 0.078 $\ell$ | 0.005 $\ell$ |
| *Butane* | *25 $\ell$* | *0.048 $\ell$* | *0.002 $\ell$* | *Toluene* | *2700 $\ell$* | *13 $\ell$* | *0.13 $\ell$* |
| *Benzene* | *85 $\ell$* | *1.3 $\ell$* | *0.041 $\ell$* | *o-Xylene* | *7500 $\ell$* | *30 $\ell$* | *0.23 $\ell$* |
| *Pentane* | *200 $\ell$* | *0.55 $\ell$* | *0.003 $\ell$* | *p-Xylene* | *11000 $\ell$* | *70 $\ell$* | *0.49 $\ell$* |
| *Ethylbenzene* | *2500 $\ell$* | *23 $\ell$* | *0.22 $\ell$* | *m-Xylene* | *11000 $\ell$* | *42 $\ell$* | *0.31 $\ell$* |
| *Hexane* | *2600 $\ell$* | *5.1 $\ell$* | *0.030 $\ell$* | *Heptane* | *11000 $\ell$* | *22 $\ell$* | *0.10 $\ell$* |

in concentration if several smoke samples (25 $\ell$) are adsorbed at 20 °C and then desorbed at 360 °C within a smaller volume (2 $\ell$, volume of the HTMC). In fact, if the contents of the TD tube are desorbed in 2 $\ell$ of gas, we expect an increase in concentration for all compounds for which the breakthrough volume at room temperature is larger than 2 $\ell$ (assuming 100% recovery). This is not the case, for example, for methane and ethane. However, these compounds can be readily detected well below the recommended exposure limits (REL) without preconcentration.

The thermal desorption tube was conditioned with 30 m$\ell$/min nitrogen (purity 5.0) at 370 °C for 90 min and then with 150 m$\ell$/min nitrogen 5.0 at 370 °C for 35 min. Then, a flow of 100 m$\ell$/min nitrogen 5.0 at 360 °C through the thermal desorption tube was used to fill the HTMC up to a pressure of 930 mbar. The HTMC was heated and kept at 150 °C. The absorption spectrum was measured (Fig. 5.11) and was found to be "clean". Two previously measured samples (samples H01e and H01f, Fig. 5.9) were flushed through the cold (room temperature) thermal desorption tube at 20–50 m$\ell$/min (total volume of the samples: less than 1 $\ell$). Since the volume of the HTMC is 2 $\ell$, no increase in concentration will occur. The desorption took place immediately afterwards with 50 m$\ell$/min nitrogen 5.0 at 360 °C through the TD tube and into the HTMC up to a final pressure of 929 mbar. The HTMC was kept at 150 °C for this measurement. The measured absorption spectrum is shown in Fig. 5.12. Formaldehyde was detected at a

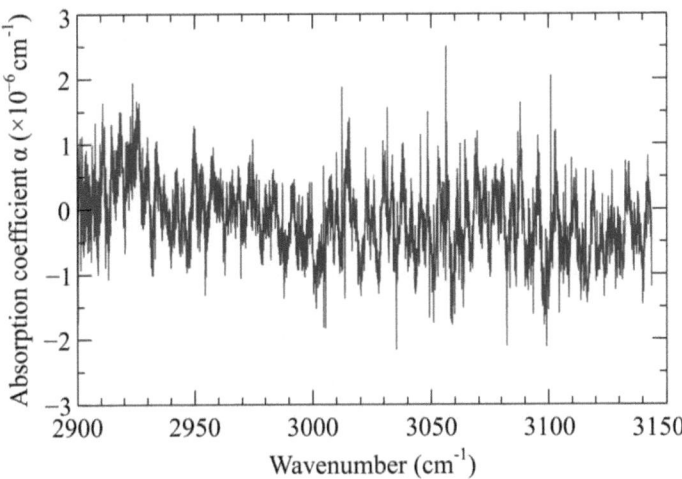

**Figure 5.11** – Absorption spectrum of nitrogen flushed through the heated (360 °C) thermal desorption tube after conditioning measured at $p = 955$ mbar and $T = 25$ °C.

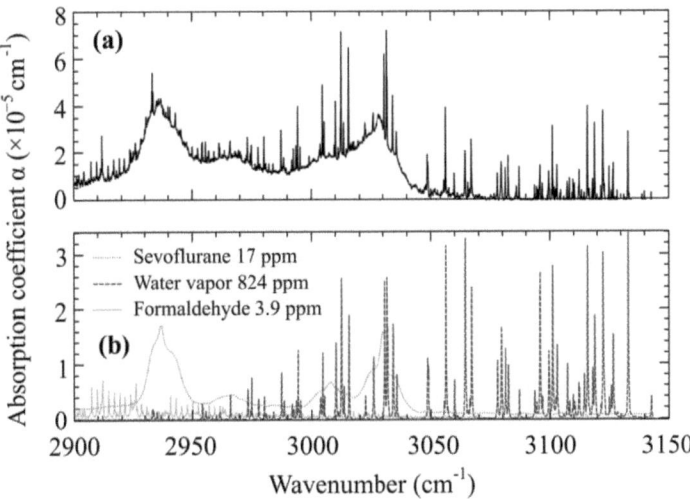

**Figure 5.12** – (a) Absorption spectrum of the thermally desorbed samples H01e and H01f measured at 150 °C and 950 mbar. (b) Identified components of (a): sevoflurane (17 ppm), water vapor (824 ppm) and formaldehyde (3.9 ppm).

relatively large concentration of 3.9 ppm. Sevoflurane is known to be unstable at high temperatures [123]. Thus, formaldehyde could be the results of a chemical process taking place either in the thermal desorption tube during desorption, or in the HTMC which was kept at 150 °C. The temperature in the thermal desorption tube is higher, but ideally all adsorbed compounds would leave the tube very quickly once nitrogen is pumped through it. The sevoflurane concentration was 17 ppm. Notice that this value is actually below the detection threshold stated in Tab. 4.2. This can be attributed to the very limited interference between the sevoflurane spectrum and the other components in the gas mixture (water vapor and formaldehyde).

To determine where formaldehyde is formed, three additional measurements were carried out. In the first one, shown in Fig. 5.13, the spectrum of sevoflurane vapor was measured with the HTMC at 150 °C. No use of TD was made: the sample was introduced into the HTMC with the device illustrated in Fig. 5.6. The formaldehyde absorption lines detected earlier (Fig. 5.12) are clearly visible. The temperature of 150 °C in the HTMC certainly contributes to the formaldehyde production. The second measurement was performed with the HTMC at 25 °C and by desorbing sevoflurane at 300 °C. The nitrogen flow through the desorption tube was initiated immediately after turning on the heating and was maintained for about 3 min after the temperature reached 300 °C. Formaldehyde formation was not observed in this case. For the third measurement, the thermal desorption tube was first heated to 300 °C and only then was the nitrogen flow

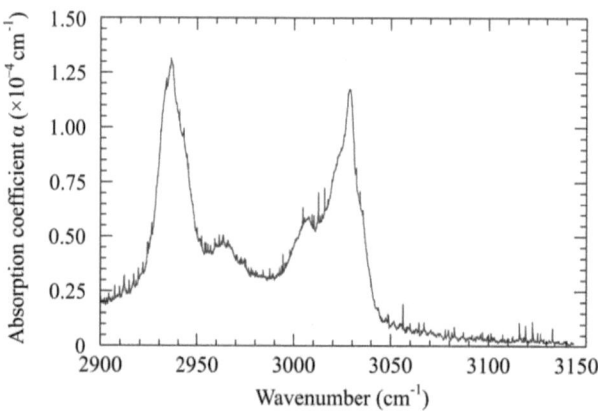

**Figure 5.13** – Absorption spectrum of sevoflurane vapor measured at 150 °C and 955 mbar. The formaldehyde absorption lines are clearly visible (cf. Fig. 5.12).

started. Formaldehyde was detected at 1.5 ppm. We can therefore gather that formaldehyde forms at temperatures of 150 °C or less, both in the HTMC and in the thermal desorption tube. In order to minimize formaldehyde production (and possibly other chemical processes), the amount of time sevoflurane (and any other adsorbed compound) remain in the hot thermal desorption tube should be minimized. Also, future measurements should be carried out at room temperature to prevent chemical processes in the HTMC.

The samples H04a–H04i (Tab. 5.4) were all collected during the same operation at USZ (total volume approximately 25 $\ell$). They were first filled into the HTMC and measured as usual. But after each measurement, instead of discarding the sample, it was pumped at a flow of 20–50 m$\ell$/min through the thermal desorption tube at room temperature. After the last sample had been measured, the contents of the thermal desorption tube were desorbed at 270 °C with 400 m$\ell$/min nitrogen 5.0. Given the initial volume of the smoke samples (25 $\ell$) and the volume after desorption (2 $\ell$) we expect an increase in concentration in all compounds retained by the absorber by a factor of 12.5. The absorption spectrum of this sample is shown in Fig. 5.14. Formaldehyde (approximately 5 ppm) and water vapor absorption lines are indicated. In view of what was said earlier about formaldehyde production in the thermal desorption tube, we can assume

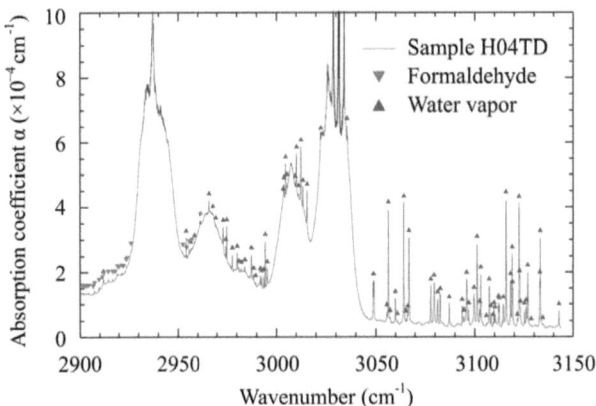

**Figure 5.14** – Absorption spectrum of sample H04TD measured at $T = 25$ °C and $p = 950$ mbar. Formaldehyde absorption lines are visible between 2900 and 2925 cm$^{-1}$. The remaining absorption lines are due to water vapor. Sevoflurane (the four broad peaks) saturates the absorption near 2940 and 3030 cm$^{-1}$.

that formaldehyde was not present in the surgical smoke samples, but was produced during the desorption. Its relatively large concentration could be explained by the large sevoflurane concentration (approx. 1000 ppm). Methane and ethane were not detected, but this is probably due to their small breakthrough volumes (0.035 $\ell$ and 0.50 $\ell$, respectively, see Tab. 5.2). Apart from sevoflurane, water vapor and formaldehyde no other compound could be detected in this preconcentrated sample.

Although the presence of sevoflurane in surgical smoke samples is interesting, it also hinders the detection of other compounds, especially those that absorb only below 3050 $cm^{-1}$, where absorption due to sevoflurane is strong (Fig. 5.7). This is particularly true for sample H04TD, which was preconcentrated with TD. The absorption is saturated near 2940 and 3030 $cm^{-1}$ and is generally strong between 2900 and 3050 $cm^{-1}$. This may limit the detection thresholds of compounds that absorb in this range to values below the ones stated in Tab. 4.2, despite the preconcentration.

What was said about the detection thresholds in Tab. 4.2 applies here as well: to systematically determine limits of detection it would be necessary to add small amounts of an undetected substance to a sample and verify at which concentration that substance becomes detectable.

The composition of all samples measured with the DFG spectrometer is given in Tab. 5.4 in Sec. 5.2.4.

## 5.2.2 Measurements with the FTIR Spectrometer

The absorption spectrum of sample H01f, shown in Fig. 5.15, was measured with an FTIR spectrometer (Bruker Optics, U.S.A., model IFS 66v) with a resolution of 0.125 $cm^{-1}$ [119]. Water vapor and carbon dioxide saturate the absorption in three regions: 1320– 1910 $cm^{-1}$ (water), 2240–2380 $cm^{-1}$ (carbon dioxide), and 3530–3960 $cm^{-1}$ (both). Outside of them, there are two regions of interest. One is around 3000 $cm^{-1}$, is accessible with the DFG spectrometer and is shown in Fig. 5.16. Notice how the signal-to-noise ratio is much better in the spectrum measured with the DFG spectrometer (Fig. 5.16b). This is due, in part, to the longer absorption pathlength (35 m versus 4 m of the FTIR). Several narrow absorption lines – water vapor in this case – are hidden in the noise of the FTIR spectrum (Fig. 5.16a). The sensitivity of the DFG spectrometer is obviously better, but with the FTIR spectrometer measurements of stronger absorption lines/bands outside of the tuning range of the DFG spectrometer may lead to a lower limit of detection. Compounds that have their largest absorption cross section between 2900 and 3150 $cm^{-1}$ (e.g., methane) can be measured more accurately with the DFG spectrometer. The other region of interest is around

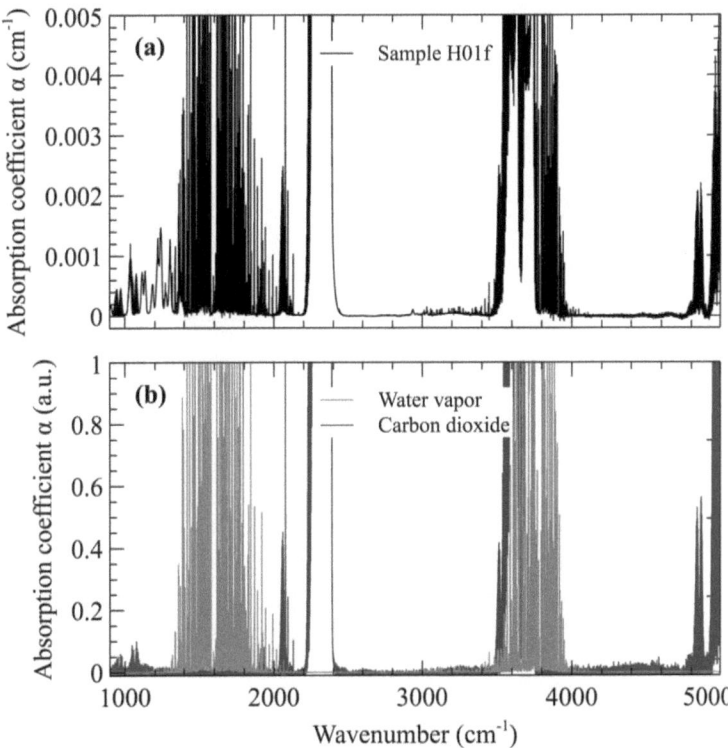

**Figure 5.15** – **(a)** FTIR absorption spectrum of sample H01f measured at ambient pressure and room temperature. **(b)** Water vapor and carbon dioxide saturate the spectrum between 1320–1910 cm$^{-1}$ (water), 2240–2380 cm$^{-1}$ (carbon dioxide), and 3530–3960 cm$^{-1}$ (both).

1000 cm$^{-1}$ (Fig. 5.17). Apart from the the 10P, 10R, 9P and 9R branches of $CO_2$ there are at least eight broad peaks. No additional compound could be detected. In particular, none of the substances revealed with FTIR spectroscopy on smoke samples produced in the lab (hydrogen cyanide, nitric and nitrous oxide, carbon monoxide and acetylene, see Sec. 4.2) were detected here. We have compared this part of the spectrum to the spectrum of sevoflurane published in Ref. [124] and it matches. The FTIR measurement thus confirms the findings obtained with the DFG spectrometer. The detection thresholds, based on a noise equivalent absorption coefficient of $\alpha = 1 \times 10^{-5}$ cm$^{-1}$ are given in Tab. 5.3.

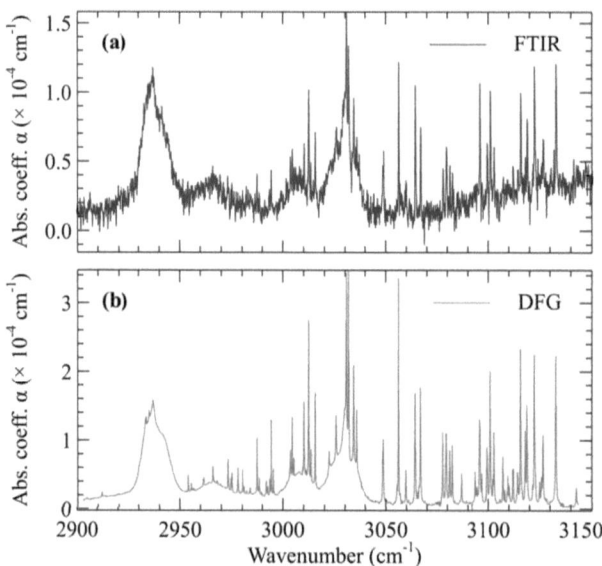

**Figure 5.16** – (a) Magnification of the spectrum in Fig. 5.15 between 2900 and 3150 cm$^{-1}$. (b) Spectrum of the same sample measured with the DFG spectrometer ($p$ = 950 mbar, $T$ = 25 °C). The broad peaks are due to sevoflurane, the narrow absorption lines to water vapor.

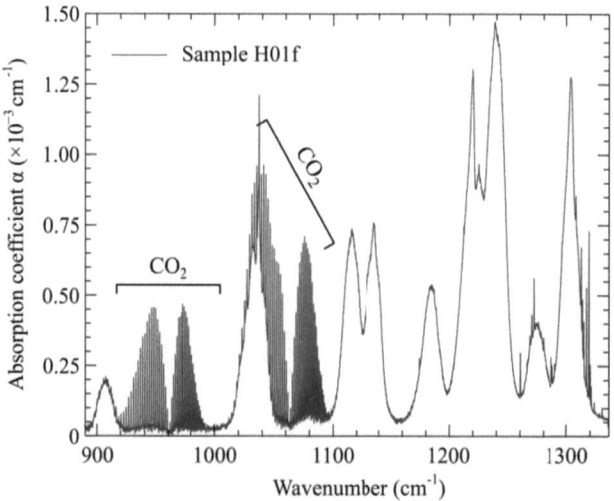

**Figure 5.17** – Magnification of the spectrum in Fig. 5.15 between 900 and 1300 cm$^{-1}$. All the broad absorption peaks are due to sevoflurane.

**Table 5.3** – Detection thresholds with the FTIR spectrometer for the five compounds detected in smoke samples produced in the lab.

| Substance | $c_{\min}$ ppm | Substance | $c_{\min}$ ppm |
|---|---|---|---|
| Carbon monoxide | 0.5 | Nitric oxide | 1.4 |
| Hydrogen cyanide | 1.2 | Nitrous oxide | 0.3 |
| Acetylene | 0.9 | | |

## 5.2.3 Measurements with the DFB Spectrometer

With the DFB spectrometer (see Sec. 2.2.2), surgical smoke samples from the last two operations (H05 and H06) were analyzed with respect to carbon monoxide (CO) and hydrogen fluoride (HF) content. A wavemeter for this wavelength was not available, but fortunately there are methane absorption lines close to the CO or HF lines that can be used for wavelength calibration. Most measurements were followed and/or preceded by a measurement of methane, so that the horizontal axis can be given in wavelength/wavenumber instead of arbitrary units. Due to these lines and to a few water vapor absorption lines, it is necessary to work at reduced pressure in order to minimize overlaps. All measurements with the DFB spectrometer were carried out at a pressure of 100 mbar and at room temperature.

**Wavelength Modulation**

In the first operation, CO was detected in only one sample at 300 ppb, while in all the samples from the second operation larger concentrations were found (Fig. 5.18). The vertical axis in Fig. 5.18 has been computed, similarly as for Fig. 2.24, with Eq. (2.75), but with $k_B$ instead of $k_A$ (see Sec. 2.5.3). The CO concentrations follow from

$$c \text{ (ppm)} = \frac{\alpha \text{ (cm}^{-1})}{\alpha_1 \text{ (cm}^{-1})} = 8.22 \times 10^6 \alpha \text{ (cm}^{-1}), \tag{5.3}$$

where $\alpha_1 = 1.22 \times 10^{-7}$ cm$^{-1}$ is the absorption coefficient for 1 ppm of CO at the peak of the absorption line for a pressure of 100 mbar and at room temperature [107]. The computed concentrations are 3.2 ppm, 2.6 ppm and 1.4 ppm. For comparison, the recommended exposure limit (REL, [16]) is 30 ppm (time-weighted average over eight hours).

In general, carbon monoxide production during laparoscopic surgery is of some concern, as several previous studies have pointed out [10, 125, 126].

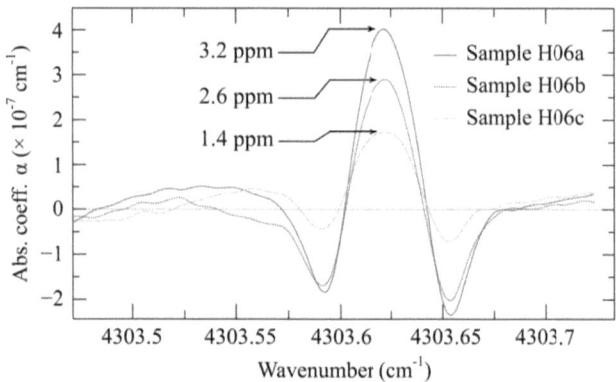

**Figure 5.18** – Carbon monoxide absorption line at 4303.62 cm$^{-1}$ in samples H06a–c.

However, reported concentrations in surgical smoke in excess of 300 ppm [125] or even 490 ppm [10] could not be confirmed here (the highest measured concentration was 3.2 ppm). The difference could be explained by the different tools employed in the cited studies.

Due to the presence of a fluorinated compound (sevoflurane) in the surgical smoke samples, we hypothesized that hydrogen fluoride could be produced during electrocautery. Unfortunately, it could not be detected unequivocally in any of the six measured samples. In Fig. 5.19 the three measurements on samples H06a–c are shown. In the first measurement (sample H06a), there is a feature that resembles the expected lineshape for second harmonic detection (Fig. 2.5) at the correct wavelength (4109.94 cm$^{-1}$). If it were a HF absorption line, the concentration could be computed, similarly to Eq. (5.3), with

$$c\,(\mathrm{ppb}) = \frac{\alpha\,(\mathrm{cm}^{-1})}{\alpha_1\,(\mathrm{cm}^{-1})} = 1.13 \times 10^7 \alpha\,(\mathrm{cm}^{-1}), \tag{5.4}$$

where $\alpha_1 = 8.82 \times 10^{-8}$ cm$^{-1}$ is the absorption coefficient for 1 ppb of HF at the peak of the absorption line for a pressure of 100 mbar and at room temperature [107]. The concentration for this line would then be approximately 40 ppt. The contrast of this possible absorption line with the surrounding fringes and noise is insufficient to make a clear statement. A concentration of 80 ppt, however, could be measured. The detection threshold given in Tab. 2.2 (110 ppt) is therefore slightly conservative in this instance. In samples H06b and H06c, HF could not be detected.

Hydrogen fluoride is very polar, so that if very small amounts are pro-

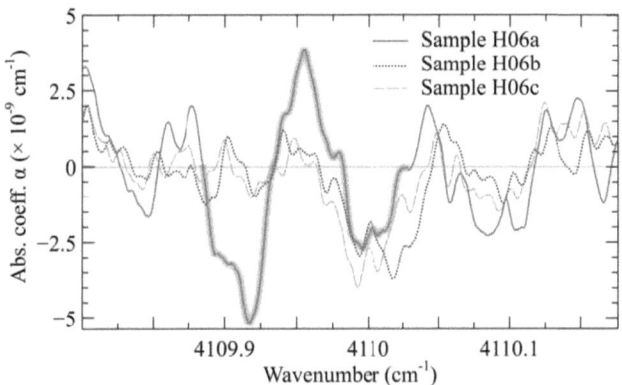

**Figure 5.19** – Absorption spectrum of samples H06a–c measured with the DFB spectrometer with diode A. The highlighted part is a possible HF absorption line (40 ppt).

duced and then transferred into a sample bag, it is very likely that HF molecules will stick to the connection tubes and to the walls of the sample bag, and may never reach the measurement cell. With the device illustrated in Fig. 5.20 we produced gas samples by creating an electric discharge with an electroknife. It is the same as the cell used previously (see Sec. 4.1) with

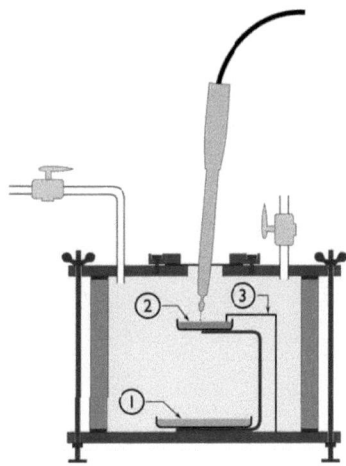

**Figure 5.20** – Cell for the production of hydrogen fluoride by electric discharge with an electroknife. (1) Liquid sevoflurane; (2) water; (3) grounding wire.

the addition of two dishes mounted on a holder inside the cell. The tube that carries the gas out of the cell is made of stainless steel. A few m$\ell$ of liquid sevoflurane were put in the lower dish and a few m$\ell$ of water in the upper dish. The cell was flushed with carbon dioxide for two minutes. All valves were closed and we waited one minute to allow sufficient time for a part of the liquid sevoflurane to evaporate. Then, the electroknife was activated while holding its tip close or in contact with the water surface to obtain an electric discharge (Fig. 5.21). Several samples were produced with different discharge durations and filled into Tedlar bags. One sample was measured with the DFG spectrometer and showed total absorption between 2900 and 3144 cm$^{-1}$. We can assume that the sevoflurane vapor concentration was extremely high ($\gg$ 1000 ppm), which is to be expected given its high vapor pressure (263 mbar). In two out of six samples measured with the DFB spectrometer hydrogen fluoride was detected along with methane. One such measurement is shown in Fig. 5.22a. The hydrogen fluoride concentration, computed with Eq. (5.4), is 280 ppt. The methane concentration is about a factor $10^4$ larger. Carbon monoxide was detected in all six measured samples at concentrations in excess of 1000 ppm. One measurement is shown in Fig. 5.22b.

In the next step, the water in the upper dish was replaced with sevoflurane, and the discharge occurred directly over the sevoflurane surface. Four samples were produced with increasingly longer discharge durations. The results are summarized in Fig. 5.23. Hydrogen fluoride could be detected in every sample, at increasingly larger concentrations. The HF concentrations computed with Eq. (5.4) are 1.1 ppb, 2.3 ppb, 3.3 ppb (Fig. 5.23a)

**Figure 5.21** – Electric discharge with an electroknife on water surface in a carbon dioxide/sevoflurane vapor atmosphere.

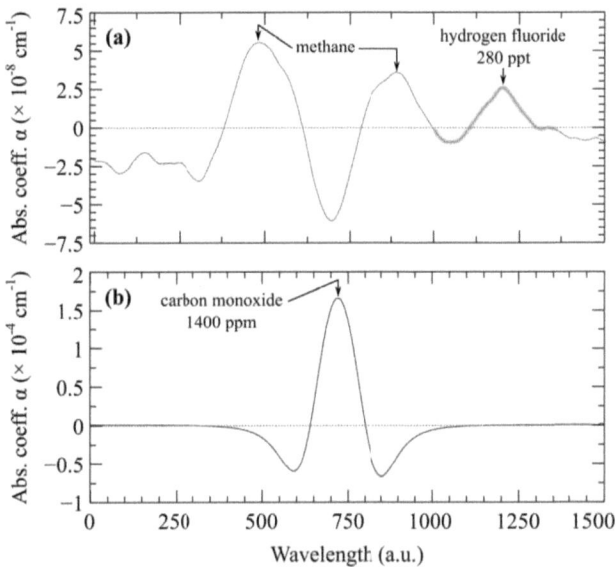

**Figure 5.22** – Wavelength modulation measurements on samples produced by creating an electric discharge in a carbon dioxide/sevoflurane vapor atmosphere. (a) Methane and hydrogen fluoride (4109.94 cm$^{-1}$, 280 ppt) absorption lines measured with diode A. (b) Carbon monoxide absorption line (4303.63 cm$^{-1}$, 1400 ppm) measured with diode B.

and 130 ppb (Fig. 5.23b). After these measurements, 10 ppm of methane were measured to confirm that the absorption lines detected previously belong to HF, which is the case, as can be seen by comparing Fig. 5.23c with Fig. 2.20a.

The recommended exposure limit (REL) for hydrogen fluoride is 1 ppm [16]. Even with an aggressive discharge in a saturated sevoflurane vapor atmosphere – both conditions which are unlikely to occur during surgery – the highest HF concentration was found to be 130 ppb when the electric discharge took place near liquid sevoflurane. Thus, given the low observed concentrations it is unlikely that hydrogen fluoride could pose a serious health hazard.

**Direct Transmission**

It was mentioned earlier (see Sec. 2.5.2), that the DFB spectrometer can also be used in direct transmission mode by switching off the fast sinusoidal wavelength modulation. Noise suppression is achieved both with a

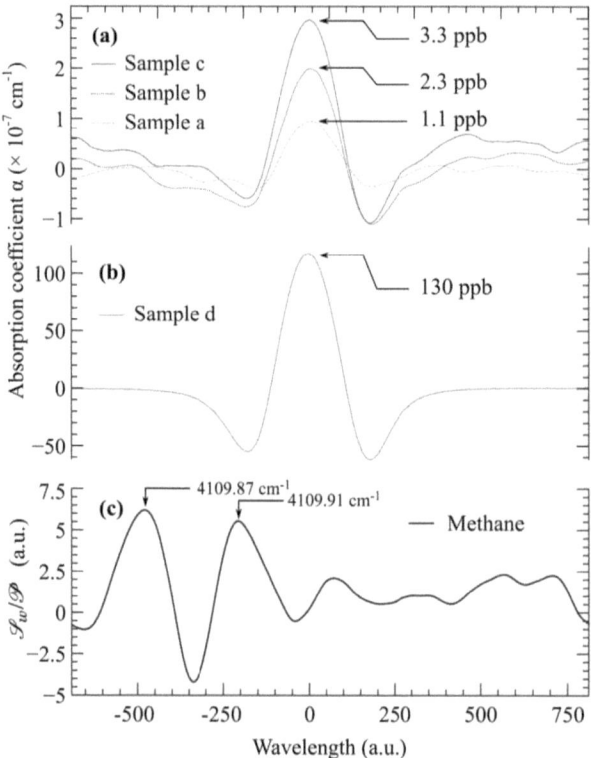

**Figure 5.23** – (a) Hydrogen fluoride absorption line at 4109.94 cm$^{-1}$ for the first three samples a–c produced by creating an electric discharge with an electroknife over (liquid) sevoflurane. (b) Sample d. (c) Methane absorption lines measured to confirm that the position of the previously measured absorption lines is that of hydrogen fluoride.

low-pass filter prior to the acquisition electronics, and by averaging several scans. The averaged scan is then low-pass-filtered again to remove high-frequency noise added by the ADC. This last step does not influence the detection bandwidth, as long as the cut-off frequency is not chosen smaller than that of the previous low-pass filter. The cut-off frequency of the low-pass filters depends on the scan speed and on the lineshape. We found that for a lineshape as the one depicted in Fig. 5.23b measured over 1 ms a cut-off frequency of 21 kHz did not significantly distort the signal. Further reduction of the detection bandwidth is achieved by averaging 60000 scans acquired during 2 minutes at a repetition rate of 1 kHz (the total acquisition

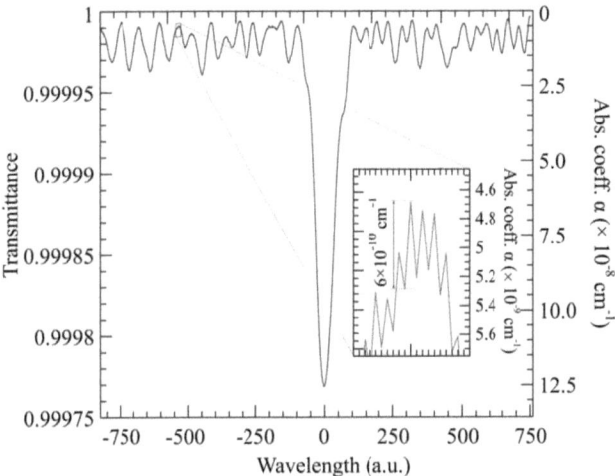

**Figure 5.24** – Carbon monoxide absorption line at 4303.62 cm$^{-1}$ measured in direct transmission. Inset: magnification to show the noise equivalent absorption coefficient: $\alpha_{\text{ne}} = 6 \times 10^{-10}$ cm$^{-1}$.

time is actually only 1 min because only every other spectrum is acquired). The total detection bandwidth (including averaging) is $B = 0.35$ Hz.

The absorption line of carbon monoxide at 4303.62 cm$^{-1}$ for a concentration of 1 ppm is shown in Fig. 5.24. Interference fringes are clearly visible and have an amplitude of approx. $10^{-8}$ cm$^{-1}$. It should be noted, that the laser diode mount and the MPC were "vibrated" (see Sec. (2.5.4)) in the same way as for the wavelength modulation measurements. Without this, the amplitude of the fringes would be much larger. The noise equivalent absorption coefficient is $\alpha_{\text{ne}} = 6 \times 10^{-10}$ cm$^{-1}$. With a detection bandwidth of $B = 0.35$ Hz the 1-Hz bandwidth-normalized noise equivalent absorption coefficient is $\alpha_{1\,\text{Hz}} = 1.0 \times 10^{-9}$ cm$^{-1}$Hz$^{-1/2}$. This value is only slightly larger than in the case with wavelength modulation ($8.5 \times 10^{-10}$ cm$^{-1}$Hz$^{-1/2}$, Tab. 2.2). This measurement shows that the same sensitivity can be achieved in direct transmission mode as in wavelength modulation mode, but if the fringes could be totally suppressed, wavelength modulation would provide a slightly better sensitivity.

## 5.2.4 Summary of the Obtained Results

The composition of all measured samples is given in Tab. 5.4. A description of the samples can be found in Tab. 5.1.

**Table 5.4** – Summary of the measured surgical smoke samples. A dash (—) indicates that the samples were not measured with the given spectrometer.

| Sample[*] | DFG Concentrations | | | | | DFB Concentrations | |
|---|---|---|---|---|---|---|---|
| | H$_2$O % | Sevo[†] ppm | CH$_4$ ppm | C$_2$H$_6$ ppm | C$_2$H$_4$ ppm | CO ppm | HF ppb |
| **H01** | | | | | | | |
| a | 0.76 | <20 | <0.1 | <0.1 | <5 | — | — |
| b | 1.1 | 450 | 0.74 | <0.1 | <5 | — | — |
| c | 0.8 | 430 | 2.2 | <0.1 | <5 | — | — |
| d | 0.27 | 180 | 0.40 | <0.1 | <5 | — | — |
| e | 0.94 | 120 | 1.1 | <0.1 | <5 | — | — |
| f | 0.59 | 150 | <0.1 | <0.1 | <5 | — | — |
| **H02** | | | | | | | |
| a | 0.72 | 290 | 0.70 | <0.1 | <5 | — | — |
| b | 0.53 | 180 | 0.45 | <0.1 | <5 | — | — |
| c | 0.65 | 160 | 0.39 | <0.1 | <5 | — | — |
| d | 0.68 | 120 | 2.2 | 0.19 | <5 | — | — |
| e | 0.58 | 240 | 5.3 | 1.1 | 6.3 | — | — |
| f | 0.58 | 300 | 9.1 | 2.0 | 10 | — | — |
| **H03** | | | | | | | |
| a | 0.98 | <20 | <0.1 | <0.1 | <5 | — | — |
| b | 0.79 | <20 | 1.1 | <0.1 | <5 | — | — |
| c | 0.68 | <20 | 1.3 | <0.1 | <5 | — | — |
| d | 0.89 | <20 | <0.1 | <0.1 | <5 | — | — |
| e | 0.83 | <20 | 1.3 | <0.1 | <5 | — | — |
| f | 0.78 | <20 | 0.45 | <0.1 | <5 | — | — |

| Sample | DFG Concentrations | | | | | DFB Concentrations | |
|---|---|---|---|---|---|---|---|
| | H$_2$O % | Sevo ppm | CH$_4$ ppm | C$_2$H$_6$ ppm | C$_2$H$_4$ ppm | CO ppm | HF ppb |
| **H04** | | | | | | | |
| a | 0.89 | 56 | <0.1 | <0.1 | <5 | — | — |
| b | 0.61 | 59 | <0.1 | <0.1 | <5 | — | — |
| c | 0.60 | 78 | <0.1 | <0.1 | <5 | — | — |
| d | 0.44 | 110 | 0.15 | <0.1 | <5 | — | — |
| e | 0.58 | 80 | 0.22 | <0.1 | <5 | — | — |
| f | 0.59 | 130 | <0.1 | <0.1 | <5 | — | — |
| g | 0.45 | 110 | <0.1 | <0.1 | <5 | — | — |
| h | 0.55 | 69 | 0.16 | <0.1 | <5 | — | — |
| i | 0.60 | 87 | 0.15 | <0.1 | <5 | — | — |
| TD | 0.57 | 630 | <0.1 | <0.1 | <5 | — | — |
| **H05** | | | | | | | |
| a | — | — | — | — | — | 0.30 | <0.11 |
| b | — | — | — | — | — | <0.25 | <0.11 |
| c | — | — | — | — | — | <0.25 | <0.11 |
| **H06** | | | | | | | |
| a | 0.47 | <20 | <0.1 | <0.1 | <5 | 3.2 | <0.11 |
| b | — | — | — | — | — | 2.6 | <0.11 |
| c | 0.99 | 250 | 34 | <0.1 | <5 | 1.4 | <0.11 |

[*] For a description of the samples see Tab. 5.1.
[†] Sevoflurane C$_4$H$_3$F$_7$O (CAS Nr. 28523-86-6).

# Chapter 6
# Conclusions and Outlook

## Conclusions

We have presented the first infrared laser spectroscopic study on surgical smoke produced during minimally invasive surgery. In a first step, animal meat tissue was cauterized with a high-frequency electroknife in a carbon dioxide atmosphere. The produced smoke samples were measured with a DFG spectrometer (see Ch. 4) and the resulting spectra were evaluated with a special algorithm (see Ch. 3). Apart from the expected water vapor, some relatively harmless hydrocarbons (methane, ethane and ethylene) were detected in most investigated samples with concentrations in the ppm range (Tab. 4.1). There was an additional contribution from one or more unknown compounds that was observed in all spectra (Fig. 4.7). Unfortunately, not even with FTIR spectroscopy could its origin be established. However, some interesting additional substances – which could not be detected with the DFG spectrometer – were found (Fig. 4.10): hydrogen cyanide, nitric and nitrous oxide, carbon monoxide and acetylene, all with concentrations in the tens and hundreds of ppm range.

In a second step, actual surgical smoke samples were obtained from the University hospital Zurich (USZ) during routine colorectal laparoscopic surgery. They were measured and analyzed in the same way as the samples produced in the lab. The broad absorption feature observed previously was not present anymore. In most samples traces of the employed anesthetic (sevoflurane) were found with concentrations between < 20 and 450 ppm. Methane was also detected in most samples at sub-ppm concentrations (Tab. 5.4). Apart from very few samples where ethane and ethylene were detected as well, methane, sevoflurane and water vapor are the only three substances needed to fully explain the measured absorption spectra (Fig. 5.5). An FTIR spectrum did not reveal any additional compounds. In particular, none of the previously detected compounds (nitric and nitrous oxide, carbon monoxide, acetylene and hydrogen cyanide) was present in surgical smoke. Preconcentration of samples with thermal desorption tubes was explored but seems problematic due to the instability of sevoflurane (formaldehyde production). Moreover, since sevoflurane itself was adsorbed, the absorption spectrum was saturated at several wavelengths. Hence, sevoflurane interferes with other possible compounds and the resulting detection thresholds may actually be lower despite the preconcentration.

Measurements with the DFB spectrometer aimed at determining carbon monoxide and hydrogen fluoride concentrations were partially successful. Carbon monoxide was identified in four out of six measured samples at concentrations ranging from 300 ppb up to 3.2 ppm (limit of detection: 250 ppb). The much larger concentrations reported in previous studies [10, 125] (above 490 ppm) could not be confirmed. This could be due to different surgical tools and/or power settings being employed. The production of hydrogen fluoride was hypothesized based on the presence of a fluorinated anesthetic (sevoflurane) in the surgical smoke. However, it could not be unequivocally detected in any of the six measured samples. With a limit of detection around 100 ppt and a recommended exposure limit (REL) of 1 ppm (10000 larger), we can affirm that hydrogen fluoride, if produced at all, does not pose a health hazard to operation room personnel. The concentration within the abdominal cavity of the patient may well be larger, but we do not have the capability of measuring *in situ*. Further measurements on gas samples produced by creating an electric discharge in a sevoflurane vapor/carbon dioxide atmosphere showed the presence of carbon monoxide (> 1000 ppm) and of hydrogen fluoride (240 ppt–110 ppb). It was also shown that very similar detection thresholds (for CO and HF) can be achieved with direct transmission, which is somewhat simpler than wavelength modulation and requires no calibration.

From the large number of substances found in previous – mostly gas-chromatographic mass-spectrometric – studies of surgical smoke (Tabs. 1.1 and 1.2), we expected a much richer composition. However, earlier studies were often carried out *in vitro* or concerned open surgery, and cannot be compared directly with the conditions and tools employed in laparoscopic surgery. It is certainly true that the limits of detection, which for the DFG spectrometer lie in the 1–100 ppm range for many substances, were not sufficiently low for the sensing of compounds other than those listed in Tab. 5.4. But one should keep in mind that a detection threshold below the recommended exposure limit of a toxic substance is unnecessary for the purpose of risk assessment. This is the case, for example, for $n$-hexane, which can be detected at 2 ppm and whose REL is 25 times larger (Tab. 4.2).

The limited sensitivity is due in part to interference with water vapor absorption lines, and in part to baseline drifts that occur between the sample and reference measurement (see Sec. 2.4.4). It was said earlier that baseline drifts are relatively unimportant when dealing with narrow absorption lines. Several interesting compounds (Tab. 1.2), however, do not have resolvable absorption line structures at ambient pressure. Measuring at reduced pressure ($p < 100$ mbar) would theoretically improve both sensitivity and selectivity of the spectrometer, but it requires a sufficiently narrow laser linewidth and a sufficiently small tuning step size. Neither of these conditions are fulfilled here (Tab. 2.1, Fig. 4.9). In the actual surgical smoke samples not only water vapor but sevoflurane as well interferes with the detection of further substances. It is probably for this reason that the preconcentrated sample produced with thermal desorption did not reveal any additional compounds.

The most interesting finding seems to be the presence of sevoflurane within surgical smoke. It is not clear by which mechanism sevoflurane reaches the abdominal cavity. The measured concentrations were large compared to the recommended exposure limit of 2 ppm [122]. Among the detected compounds, and taking into consideration the average concentrations, this was the most dangerous detected substance for the operation room personnel.

# Outlook

The multitude of chemical species detected in earlier examinations raises the question of whether the composition of the samples investigated during this study was qualitatively different, or if the concentrations were so low that they could not be measured. The sensitivity was limited both by the properties of the spectrometer and by interference due to water vapor and sevoflurane. Measurements at lower pressure ($p < 100$ mbar) would reduce the overlap of absorption lines, thus increasing sensitivity and selectivity. For this, a narrow linewidth laser source, capable of tuning in steps much smaller than the linewidth, is required. Distributed feedback diode and quantum cascade lasers are suitable for this task, as has been shown for hydrogen fluoride and carbon monoxide. But the increased sensitivity and selectivity comes at the price of a much smaller tuning range: the set of substances detectable with a single laser source is limited. For an exhaustive analysis multiple laser sources – each one aimed at one or few specific compounds – would be needed. Wavelength regions that should be avoided because of strong carbon dioxide and/or water vapor absorption are 1320–1910 cm$^{-1}$, 2240–2380 cm$^{-1}$ and 3530–3960 cm$^{-1}$. Sevoflurane has several strong absorption bands near 10 μm; hence, detection of additional substances at this wavelength could also be problematic. The absorption near 3000 cm$^{-1}$ is weaker, but still sufficient to reduce the sensitivity to other substances.

The large tuning range of OPOs would provide access to numerous compounds with a single laser source, but due to their typically small *continuous* tuning range, the detection of a set of compounds would require several measurements with manual adjustments between each. This would not be suitable for routine analysis of surgical smoke, but might provide sufficient sensitivity.

Some benefit can also be expected from a carefully selected sample preparation procedure. Water and carbon dioxide removal would "clean" the infrared spectrum of strong absorption lines in the three regions cited above. This would allow, for instance, the detection of aldehydes (e.g., formaldehyde, furfural), which manifest a strong absorption around 1700 cm$^{-1}$.

Finally, it would be interesting to explore the relationship, if there is one, between the composition of surgical smoke and the type of cauterized tissue.

# Bibliography

[1] M. Baggish, B. Poiesz, D. Joret, P. Williamson, and A. Rebai: Presence of Human Immunodeficiency Virus DNA in Laser Smoke, *Lasers Surg. Med.* **11**, 197–203 (1991).

[2] P. Hallmo and O. Naess: Laryngeal Papillomatosis with Papilloma Virus DNA Contracted by a Laser Surgeon, *Em. Arch. Otorhinolaryngol.* **248**, 425–427 (1991).

[3] L. Calero and T. Brusis: Laryngeal papillomatosis - First recognition in Germany as an occupational disease in an operating room nurse, *Laryngo-rhino-otologie* **82**, 790–793 (2003).

[4] J. M. Garden, M. K. O'Banion, A. D. Bakus, and C. Olson: Viral Disease Transmitted by Laser-Generated Plume (Aerosol), *Arch. Dermatol.* **138**, 1303–1307 (2002).

[5] B. Ziegler, C. Thomas, T. Meier, R. Muller, T. Fliedner, and L. Weber: Generation of Infectious Retrovirus Aerosol through Medical Laser Irradiation, *Lasers Surg. Med.* **22**, 37–41 (1998).

[6] D. Jewett, P. Heinsohn, C. Bennett, A. Rosen, and C. Neuilly: Blood-Containing Aerosols Generated by Surgical Techniques - a Possible Infectious Hazard, *American Industrial Hygiene Association Journal* **53**, 228–231 (1992).

[7] P. Heinsohn and D. Jewett: Exposure to Blood-Containing Aerosols in the Operating-Room - a Preliminary-Study, *American Industrial Hygiene Association Journal* **54**, 446–453 (1993).

[8] K. J. Weld, S. Dryer, C. D. Ames, K. Cho, C. Hogan, M. Lee, P. Biswas, and J. Landman: Analysis of Surgical Smoke Produced by Various Energy-Based Instruments and Effect on Laparoscopic Visibility, *Journal of Endourology* **21**, 347–351 (2007).

[9] A. R. Moot, K. M. Ledingham, P. F. Wilson, S. T. Senthilmohan, D. R. Lewis, J. Roake, and R. Allardyce: Composition of Volatile Organic Compounds in Diathermy Plume as Detected by Selected Ion Flow Tube Mass Spectrometry, *ANZ J. Surg.* **77**, 20–23 (2007).

[10] R. Weston, R. N. Stephenson, P. W. Kutarski, and N. J. Parr: Chemical Composition of Gases Surgeons Are Exposed to During Endoscopic Urological Resections, *Urology* **74**, 1152–1154 (2009).

[11] O. S. Al Sahaf, I. Vega-Carrascal, F. O. Cunningham, J. P. McGrath, and F. J. Bloomfield: Chemical Composition of Smoke Produced by High-Frequency Electrosurgery, *Irish Journal of Medical Science* **176**, 229–232 (2007).

[12] Y. J. Chung, S. K. Lee, S. H. Han, C. Zhao, M. K. Kim, S. C. Park, and J. K. Park: Harmful gases including carcinogens produced during transurethral resection of the prostate and vaporization, *International Journal of Urology* **17**, 944–949 (2010).

[13] Y. Tomita, S. Mihashi, K. Nagata, S. Ueda, M. Fujiki, M. Hirano, and T. Hirohata: Mutagenicity of Smoke Condensates Induced by $CO_2$-Laser Irradiation and Electrocauterization, *Mutat. Res.* **89**, 145–149 (1981).

[14] D. Galbraith, S. A. Gross, and D. Paustenbach: Benzene and Human Health: A Historical Review and Appraisal of Associations with Various Diseases, *Critical Reviews in Toxicology* **40**, 1–46 (2010).

[15] L. Lehman-McKeeman: Paracelsus and Formaldehyde 2010: The Dose to the Target Organ Makes the Poison, *Toxicological Sciences* **116**, 361–363 (2010).

[16] SUVA: *Grenzwerte Am Arbeitsplatz*, SUVA, Lucerne, Switzerland (2009).

[17] K. Nakajima, J. Milsom, and B. Böhm: Equipment and Instrumentation, in J. Milsom, B. Böhm, and K. Nakajima, editors, *Laparoscopic Colorectal Surgery*, chapter 2, pages 10–29, Springer, Berlin, 2nd edition (2006).

[18] J. Kennedy, P. Stranahan, K. Taylor, and J. Chandler: High-Burst-Strength, Feedback-Controlled Bipolar Vessel Sealing, *Surg. Endosc.* **12**, 876–878 (1998).

[19] T. Menes and H. Spivak: Laparoscopy - Searching for the Proper Insufflation Gas, *Surg. Endosc.* **14**, 1050–1056 (2000).

[20] G. W. Gokel: *Dean's Handbook of Organic Chemistry*, McGraw-Hill, 2nd edition (2004).

[21] D. L. Pavia, G. M. Lampman, G. S. Kriz, and J. R. Vyvyan: *Introduction to Spectroscopy*, Brooks/Cole, 4th edition (2009).

[22] W. D. Perkins: Fourier Transform-Infrared Spectroscopy: Part I. Instrumentation, *J. Chem. Educ.* **63**, A5–A10 (1986).

[23] M. W. Sigrist: Air Monitoring by Laser Photoacoustic Spectroscopy, in M. W. Sigrist, editor, *Air Monitoring by Spectroscopic Techniques*, volume 127 of *Chemical Analysis*, chapter 4, pages 163–238, John Wiley & Sons (1994).

[24] R. Bartlome, M. Baer, and M. W. Sigrist: High-Temperature Multipass Cell for Infrared Spectroscopy of Heated Gases and Vapors, *Rev. Sci. Instrum.* **78**, 013110 (2007).

[25] R. Bartlome, J. M. Rey, and M. W. Sigrist: Vapor Phase Infrared Laser Spectroscopy: From Gas Sensing to Forensic Urinalysis, *Anal. Chem.* **80**, 5334–5341 (2008).

[26] M. W. Sigrist, R. Bartlome, D. Marinov, J. M. Rey, D. E. Vogler, and H. Waechter: Trace Gas Monitoring with Infrared Laser-Based Detection Schemes, *Appl. Phys. B* **90**, 289–300 (2008).

[27] H. Wächter and M. W. Sigrist: Mid-Infrared Laser Spectroscopic Determination of Isotope Ratios of $N_2O$ at Trace Levels Using Wavelength Modulation and Balanced Path Length Detection, *Appl. Phys. B* **87**, 539–546 (2007).

[28] H. Wächter, J. Mohn, B. Tuzson, L. Emmenegger, and M. W. Sigrist: Determination of $N_2O$ Isotopomers with Quantum Cascade Laser Based Absorption Spectroscopy, *Opt. Express* **16**, 9239–9244 (2008).

[29] G. Wysocki, R. Lewicki, R. F. Curl, F. K. Tittel, L. Diehl, F. Capasso, M. Troccoli, G. Hofler, D. Bour, S. Corzine, R. Maulini, M. Giovannini, and J. Faist: Widely Tunable Mode-Hop Free External Cavity Quantum Cascade Lasers for High Resolution Spectroscopy and Chemical Sensing, *Appl. Phys. B* **92**, 305–311 (2008).

[30] D. S. Bomse, D. C. Hovde, S.-J. Chen, and J. A. Silver: Early Fire Sensing Using Near-IR Diode Laser Spectroscopy, *Proc. SPIE* **4817**, 73–81 (2003).

[31] S. Schilt, A. A. Kosterev, and F. K. Tittel: Performance Evaluation of a Near Infrared QEPAS Based Ethylene Sensor, *Appl. Phys. B* **95**, 813–824 (2009).

[32] W. Francke, O. Fleck, D.-L. Mihalache, and W. Wöllmer: Identification of Volatile Compounds Released from Biological Tissue during $CO_2$ Laser Treatment, *Proc. SPIE* **2323**, 423–431 (1995).

[33] C. Hensman, D. Baty, R. Willis, and A. Cuschieri: Chemical Composition of Smoke Produced by High-Frequency Electrosurgery in a Closed Gaseous Environment - an in Vitro Study, *Surg. Endosc.* **12**, 1017–1019 (1998).

[34] K. Desinger, W. Wäsche, and H. Albrecht: Comparative Investigation of Pyrolysis Products Due to Laser and RF Surgery with Regard to the Latest Device Development, *Proc. SPIE* **2624**, 234–239 (1996).

[35] W. Wäsche, H. Albrecht, and G. J. Mueller: Determination of Temperature Dependence of the Production of Volatile Organic Compounds (VOCs) during the Vaporization of Tissue Using Nd:YAG Laser, $CO_2$ Laser, and Electrosurgery Devices, *Proc. SPIE* **2323**, 393–399 (1995).

[36] M. Spleiss and L. W. Weber: Medium- and low-volatile organic compounds generated by laser tissue interaction, *Proc. SPIE* **2923**, 168–177 (1996).

[37] M. Spleiss, L. Weber, T. Meier, and B. Treffler: Identification and Quantification of Selected Chemicals in Laser Pyrolysis Products of Mammalian Tissues, *Proc. SPIE* **2323**, 409–422 (1995).

[38] H. Albrecht, R. Hagemann, W. Wäsche, G. Wagner, and G. J. Mueller: Volatile Organic Components in Laser and Electrosurgery Plume, *Proc. SPIE* **2077**, 310–317 (1994).

[39] L. W. Weber and M. Spleiss: Estimation of Risks by Chemicals Produced during Laser Pyrolysis of Tissues, *Proc. SPIE* **2323**, 464–471 (1995).

[40] H. Albrecht, W. Wäsche, and G. J. Mueller: Assessment of the Risk Potential of Pyrolysis Products in Plume Produced during Laser Treatment under OR Conditions, *Proc. SPIE* **2323**, 455–463 (1995).

[41] W. Barrett and S. Garber: Surgical Smoke - a Review of the Literature - is this Just a Lot of Hot Air?, *Surg. Endosc.* **17**, 979–987 (2003).

[42] W. Wäsche and H. Albrecht: Investigation of the Distribution of Aerosoles and VOC in Plume Produced during Laser Treatment under OR Conditions, *Proc. SPIE* **2624**, 270–275 (1996).

[43] H.-J. Weigmann, J. Lademann, H. Meffert, and W. Sterry: Permanent Gases and Highly Volatile Organic Compounds in Laser Plume, *Proc. SPIE* **2923**, 164–167 (1996).

[44] H.-J. Weigmann, J. Lademann, and J. Liebetruth: Characterization of Laser-Tissue Interaction by Laser Plume Species, *Proc. SPIE* **2077**, 264–269 (1994).

[45] P. Sagar, A. Meagher, S. Sobczak, and B. Wolff: Chemical Composition and Potential Hazards of Electrocautery Smoke, *Br. J. Surg.* **83**, 1792 (1996).

[46] J. M. Rey, D. Schramm, D. Hahnloser, D. Marinov, and M. W. Sigrist: Spectroscopic Investigation of Volatile Compounds Produced during Thermal and Radiofrequency Bipolar Cautery on Porcine Liver, *Meas. Sci. Technol.* **19**, 075602 (2008).

[47] R. Hollmann, C. Hort, E. Kammer, M. Nägele, M. W. Sigrist, and C. Meuli-Simmen: Smoke in the Operating Theatre: An Unregarded Source of Danger, *Plastic Reconstr. Surg.* **114**, 458–463 (2004).

[48] J. Wu, D. Luttmann, T. Meininger, and N. Soper: Production and Systemic Absorption of Toxic Byproducts of Tissue Combustion during Laparoscopic Surgery, *Surg. Endosc.* **11**, 1075–1079 (1997).

[49] J. Silver: Frequency-Modulation Spectroscopy for Trace Species Detection – Theory and Comparison among Experimental Methods, *Appl. Opt.* **31**, 707–717 (1992).

[50] G. Berden and R. Engeln, editors: *Cavity Ring-Down Spectroscopy – Techniques and Applications*, Wiley (2009).

[51] A. Rosencwaig: *Photoacoustics and Photoacoustic Spectroscopy*, Wiley (1990).

[52] S. Kogan: *Electronic Noise and Fluctuations in Solids*, Cambridge University Press, Cambridge, U.K. (1996).

[53] B. Widrow and I. Kollár: *Quantization Noise: Roundoff Error in Digital Computation, Signal Processing, Control, and Communications*, Cambridge University Press, Cambridge, U.K. (2008).

[54] D. Herriott, H. Kogelnik, and R. Kompfner: Off-Axis Paths in Spherical Mirror Interferometers, *Appl. Opt.* **3**, 523–526 (1964).

[55] L. Rothman, D. Jacquemart, A. Barbe, D. Benner, M. Birk, L. Brown, M. Carleer, C. Chackerian, K. Chance, L. Coudert, V. Dana, V. Devi, J. Flaud, R. Gamache, A. Goldman, J. Hartmann, K. Jucks, A. Maki, J. Mandin, S. Massie, J. Orphal, A. Perrin, C. Rinsland, M. Smith, J. Tennyson, R. Tolchenov, R. Toth, J. V. Auwera, P. Varanasi, and G. Wagner: The HITRAN 2004 Molecular Spectroscopic Database, *J. Quant. Spectrosc. Radiat. Transfer* **96**, 139–204 (2005).

[56] R. Boyd: *Nonlinear Optics*, Academic Press, 3rd edition (2008).

[57] C. Fischer and M. W. Sigrist: Mid-IR Difference Frequency Generation, in *Solid-State Mid-Infrared Laser Sources*, volume 89 of *Topics Appl. Phys.*, pages 97–140, Springer, Berlin, Heidelberg (2003).

[58] E. Lippert, S. Nicolas, G. Arisholm, K. Stenersen, and G. Rustad: Midinfrared Laser Source with High Power and Beam Quality, *Appl. Opt.* **45**, 3839–3845 (2006).

[59] M. E. Klein, P. Gross, K.-J. Boller, M. Auerbach, P. Wessels, and C. Fallnich: Rapidly Tunable Continuous-Wave Optical Parametric Oscillator Pumped by a Fiber Laser, *Opt. Lett.* **28**, 920–922 (2003).

[60] M. Vainio, J. Peltola, S. Persijn, F. J. M. Harren, and L. Halonen: Singly Resonant Cw Opo with Simple Wavelength Tuning, *Opt. Express* **16**, 11141–11146 (2008).

[61] A. Berrou, M. Raybaut, A. Godard, and M. Lefebvre: High-Resolution Photoacoustic and Direct Absorption Spectroscopy of Main Greenhouse Gases by Use of a Pulsed Entangled Cavity Doubly Resonant Opo, *Appl. Phys. B* **98**, 217–230 (2010).

[62] R. Byer and R. Herbst: Parametric Oscillation and Mixing, in Y. Shen, editor, *Nonlinear Infrared Generation*, volume 16 of *Topics Appl. Phys.*, pages 81–137, Springer, Berlin, Heidelberg (1977).

[63] J. Armstrong, N. Bloembergen, J. Ducuing, and P. Pershan: Interactions between Light Waves in a Nonlinear Dielectric, *Phys. Rev.* **127**, 1918–1939 (1962).

[64] D. Roberts: Simplified Characterization of Uniaxial and Biaxial Nonlinear Optical-Crystals - a Plea for Standardization of Nomenclature and Conventions, *IEEE J. Quantum Elect.* **28**, 2057–2074 (1992).

[65] V. Dmitriev, G. Gurzadyan, and D. Nikogosyan: *Handbook of Nonlinear Optical Crystals*, Springer, Berlin, Heidelberg (1993).

[66] M. Yamada, N. Nada, M. Saitoh, and K. Watanabe: First-Order Quasi-Phase Matched $LiNbO_3$ Waveguide Periodically Poled by Applying an External Field for Efficient Blue Second-Harmonic Generation, *Appl. Phys. Lett.* **62**, 435–436 (1993).

[67] E. Lim, M. Fejer, and R. Byer: Second-Harmonic Generation of Green Light in Periodically Poled Planar Lithium-Niobate Wave-Guide, *Electronics Letters* **25**, 174–175 (1989).

[68] J. Webjorn, F. Laurell, and G. Arvidsson: Blue-Light Generated by Frequency Doubling of Laser Diode Light in a Lithium-Niobate Channel Wave-Guide, *IEEE Photonics Technology Letters* **1**, 316–318 (1989).

[69] J. Faist, F. Capasso, D. Sivco, C. Sirtori, A. Hutchinson, and A. Cho: Quantum Cascade Laser, *Science* **264**, 553–556 (1994).

[70] R. Bruggemann, M. Petri, H. Fischer, D. Mauer, D. Reinert, and W. Urban: Computer-Controlled Diode-Laser Spectrometer with a Helium Evaporation Cryostat and Spectroscopy of the $\nu_3$ Vibration of NCO, *Appl. Phys. B* **48**, 105–110 (1989).

[71] S. Y. Zhang, D. G. Revin, J. W. Cockburn, K. Kennedy, A. B. Krysa, and M. Hopkinson: $\lambda \sim 3.1$ μm Room Temperature InGaAs/AlAsSb/InP Quantum Cascade Lasers, *Appl. Phys. Lett.* **94**, 031106-1–031106-3 (2009).

[72] J. Faist, D. Hofstetter, M. Beck, T. Aellen, M. Rochat, and S. Blaser: Bound-to-Continuum and Two-Phonon Resonance Quantum-Cascade Lasers for High Duty Cycle, High-Temperature Operation, *IEEE J. Quantum Elect.* **38**, 533–546 (2002).

[73] Y. Nakano and K. Tada: Analysis, Design, and Fabrication of GaAlAs/GaAs DFB Lasers with Modulated Stripe Width Structure for Complete Single Longitudinal Mode Oscillation, *IEEE J. Quantum Elect.* **24**, 2017–2033 (1988).

[74] Y. Sin and N. Presser: Tunable InGaAsP/InP DFB Lasers at 1.3 µm Integrated with Pt Thin Film Heaters Deposited by Focused Ion Beam, *Electronics Letters* **39**, 1823–1825 (2003).

[75] S. Kassi, M. Chenevier, L. Gianfrani, A. Salhi, Y. Rouillard, A. Ouvrard, and D. Romanini: Looking into the Volcano with a Mid-IR DFB Diode Laser and Cavity Enhanced Absorption Spectroscopy, *Opt. Express* **14**, 11442–11452 (2006).

[76] L. Naehle, S. Belahsene, M. von Edlinger, M. Fischer, G. Boissier, P. Grech, G. Narcy, A. Vicet, Y. Rouillard, J. Koeth, and L. Worschech: Continuous-Wave Operation of Type-I Quantum Well DFB Laser Diodes Emitting in 3.4 µm Wavelength Range around Room Temperature, *Electronics Letters* **47** (2011).

[77] J. L. Bradshaw, J. D. Bruno, J. T. Pham, D. E. Wortman, S. Zhang, and S. R. J. Brueck: Single-Longitudinal-Mode Emission from Interband Cascade DFB Laser with a Grating Fabricated by Interferometric Lithography, *IEE Proc.* **150**, 288–292 (2003).

[78] I. Vurgaftman, C. L. Canedy, C. S. Kim, M. Kim, W. W. Bewley, J. R. Lindle, J. Abell, and J. R. Meyer: Mid-Infrared Interband Cascade Lasers Operating at Ambient Temperatures, *New J Phys* **11**, 125015 (2009).

[79] K. R. Parameswaran, D. I. Rosen, M. G. Allen, A. M. Ganz, and T. H. Risby: Off-Axis Integrated Cavity Output Spectroscopy with a Mid-Infrared Interband Cascade Laser for Real-Time Breath Ethane Measurements, *Appl. Opt.* **48**, B73–B79 (2009).

[80] M. G. Littman and H. J. Metcalf: Spectrally Narrow Pulsed Dye-Laser without Beam Expander, *Appl. Opt.* **17**, 2224–2227 (1978).

[81] W. Zeller, L. Naehle, P. Fuchs, F. Gerschuetz, L. Hildebrandt, and J. Koeth: DFB Lasers Between 760 nm and 16 µm for Sensing Applications, *Sensors* **10**, 2492–2510 (2010).

[82] Santec: http://www.santec.com/en/products/instruments/tsl-510.

[83] A. Hugi, R. Terazzi, Y. Bonetti, A. Wittmann, M. Fischer, M. Beck, J. Faist, and E. Gini: External Cavity Quantum Cascade Laser Tunable from 7.6 to 11.4 µm, *Appl. Phys. Lett.* **95**, 061103-1–061103-3 (2009).

[84] M. Brandstetter, A. Genner, K. Anic, and B. Lendl: Tunable External Cavity Quantum Cascade Laser for the Simultaneous Determination of Glucose and Lactate in Aqueous Phase, *Analyst* **135**, 3260–3265 (2010).

[85] A. Hugi, R. Maulini, and J. Faist: External Cavity Quantum Cascade Laser, *Semicond. Sci. Tech.* **25**, 083001 (2010).

[86] nanoplus: http://www.nanoplus.com/content/view/64/.

[87] C. Tang and J. Telle: Laser Modulation Spectroscopy of Solids, *J. Appl. Phys.* **45**, 4503–4505 (1974).

[88] E. Moses and C. Tang: High-Sensitivity Laser Wavelength-Modulation Spectroscopy, *Opt. Lett.* **1**, 115–117 (1977).

[89] D. S. Bomse, A. C. Stanton, and J. A. Silver: Frequency-Modulation and Wavelength Modulation Spectroscopies – Comparison of Experimental Methods Using a Lead-Salt Diode-Laser, *Appl. Opt.* **31**, 718–731 (1992).

[90] G. Bjorklund: Frequency-Modulation Spectroscopy – New Method for Measuring Weak Absorptions and Dispersions, *Opt. Lett.* **5**, 15–17 (1980).

[91] S. Schilt, L. Thevenaz, and P. Robert: Wavelength Modulation Spectroscopy: Combined Frequency and Intensity Laser Modulation, *Appl. Opt.* **42**, 6728–6738 (2003).

[92] J. Reid and D. Labrie: Second-Harmonic Detection with Tunable Diode-Lasers - Comparison of Experiment and Theory, *Appl. Phys. B* **26**, 203–210 (1981).

[93] L. Krause: Effective Quantization by Averaging and Dithering, *Measurement* **39**, 681–694 (2006).

[94] J. Potzick: Noise Averaging and Measurement Resolution (or 'a Little Noise is a Good Thing'), *Rev. Sci. Instrum.* **70**, 2038–2040 (1999).

[95] B. Lins, P. Zinn, R. Engelbrecht, and B. Schmauss: Simulation-Based Comparison of Noise Effects in Wavelength Modulation Spectroscopy and Direct Absorption TDLAS, *Appl. Phys. B* **100**, 367–376 (2010).

[96] J. Silver and A. Stanton: Optical Interference Fringe Reduction in Laser-Absorption Experiments, *Appl. Opt.* **27**, 1914–1916 (1988).

[97] M. Asobe, O. Tadanaga, T. Yanagawa, H. Itoh, and H. Suzuki: Reducing Photorefractive Effect in Periodically Poled ZnO- and MgO-Doped LiNbO$_3$ Wavelength Converters, *Appl. Phys. Lett.* **78**, 3163–3165 (2001).

[98] Y. F. Kong, J. K. Wen, and H. F. Wang: New Doped Lithium Niobate Crystal with High Resistance to Photorefraction – LiNbO$_3$:In, *Appl. Phys. Lett.* **66**, 280–281 (1995).

[99] M. Gianella and M. W. Sigrist: Automated Broad Tuning of Difference Frequency Sources for Spectroscopic Studies, *Appl. Opt.* **50**, A11–A19 (2011).

[100] D. Halmer, G. von Basum, M. Horstjann, P. Hering, and M. Mürtz: Time Resolved Simultaneous Detection of (NO)-N-14 and (NO)-N-15 Via Mid-Infrared Cavity Leak-out Spectroscopy, *Isot. Environ. Health Study* **41**, 303–311 (2005).

[101] F. M. Schmidt, O. Vaittinen, M. Metsala, P. Kraus, and L. Halonen: Direct Detection of Acetylene in Air by Continuous Wave Cavity Ring-Down Spectroscopy, *Appl. Phys. B* **101**, 671–682 (2010).

[102] A. A. Kosterev, A. L. Malinovsky, F. K. Tittel, C. Gmachl, F. Capasso, D. L. Sivco, J. N. Baillargeon, A. L. Hutchinson, and A. Y. Cho: Cavity Ringdown Spectroscopic Detection of Nitric Oxide with a Continuous-Wave Quantum-Cascade Laser, *Appl. Opt.* **40**, 5522–5529 (2001).

[103] S. M. Cristescu, S. T. Persijn, S. T. L. Hekkert, and F. J. M. Harren: Laser-Based Systems for Trace Gas Detection in Life Sciences, *Appl. Phys. B* **92**, 343–349 (2008).

[104] B. W. M. Moeskops, S. M. Cristescu, and F. J. M. Harren: Sub-part-per-billion monitoring of nitric oxide by use of wavelength modulation spectroscopy in combination with a thermoelectrically cooled, continuous-wave quantum cascade laser, *Opt. Lett.* **31**, 823–825 (2006).

[105] R. Grilli, L. Ciaffoni, G. Hancock, R. Peverall, G. A. D. Ritchie, and A. J. Orr-Ewing: Mid-Infrared Ethene Detection Using Difference Frequency Generation in a Quasi-Phase-Matched Linbo3 Waveguide, *Appl. Opt.* **48**, 5696–5703 (2009).

[106] A. Foltynowicz, F. M. Schmidt, W. Ma, and O. Axner: Noise-Immune Cavity-Enhanced Optical Heterodyne Molecular Spectroscopy: Current Status and Future Potential, *Appl. Phys. B* **92**, 313–326 (2008).

[107] L. Rothman, I. Gordon, A. Barbe, D. Benner, P. Bernath, M. Birk, V. Boudon, L. Brown, A. Campargue, J.-P. Champion, K. Chance, L. Coudert, V. Dana, V. Devi, S. Fally, J.-M. Flaud, R. Gamache, A. Goldman, D. Jacquemart, I. Kleiner, N. Lacome, W. Lafferty, J.-Y. Mandin, S. Massie, S. Mikhailenko, C. Miller, N. Moazzen-Ahmadi, O. Naumenko, A. Nikitin, J. Orphal, V. Perevalov, A. Perrin, A. Predoi-Cross, C. Rinsland, M. Rotger, M. Simeckova, M. Smith, K. Sung, S. Tashkun, J. Tennyson, R. Toth, A. Vandaele, and J. V. Auwera: The HITRAN 2008 Molecular Spectroscopic Database, *J. Quant. Spectrosc. Radiat. Transfer* **110**, 533–572 (2009).

[108] P. Werle, R. Mücke, and F. Slemr: The Limits of Signal Averaging in Atmospheric Trace-Gas Monitoring by Tunable Diode-Laser Absorption-Spectroscopy (TDLAS), *Appl. Phys. B* **57**, 131–139 (1993).

[109] S. W. Sharpe, T. J. Johnson, R. L. Sams, P. M. Chu, G. C. Rhoderick, and P. A. Johnson: Gas-Phase Databases for Quantitative Infrared Spectroscopy, *Appl. Spectrosc.* **58**, 1452–1461 (2004).

[110] M. Turk and A. Pentland: Eigenfaces for Recognition, *Journal of Cognitive Neuroscience* **3**, 71–86 (1991).

[111] I. Jolliffe: *Principal Component Analysis*, Springer Series in Statistics, Springer, Berlin, Heidelberg, 2nd edition (2002).

[112] M. Gianella and M. W. Sigrist: Improved Algorithm for Quantitative Analyses of Infrared Spectra of Multicomponent Gas Mixtures with Unknown Compositions, *Appl. Spectrosc.* **63**, 338–343 (2009).

[113] M. R. Nyden: Computer-Assisted Spectroscopic Analysis Using Orthonormalized Reference Spectra. Part I: Application to Mixtures, *Appl. Spectrosc.* **40**, 868–871 (1986).

[114] S.-C. Lo and C. W. Brown: Infrared Spectral Search for Mixtures in Large-Size Libraries, *Appl. Spectrosc.* **45**, 1628–1632 (1991).

[115] M. Gianella and M. W. Sigrist: Infrared Spectroscopy on Smoke Produced by Cauterization of Animal Tissue, *Sensors* **10**, 2694–2708 (2010).

[116] J. Cousin, W. Chen, D. Bigourd, M. Fourmentin, and S. Kassi: Telecom-Grade Fiber Laser-Based Difference-Frequency Generation and ppb-Level Detection of Benzene Vapor in Air around 3 µm, *Appl. Phys. B* **97**, 919–929 (2009).

[117] J. W. Milsom, B. Böhm, and K. Nakajima, editors: *Laparoscopic colorectal surgery*, Springer Science & Business, 2nd edition (2006).

[118] M. M. L. Steeghs, S. M. Cristescu, and F. J. M. Harren: The Suitability of Tedlar Bags for Breath Sampling in Medical Diagnostic Research, *Physiological Measurement* **28**, 73–84 (2007).

[119] M. W. Sigrist, R. Bartlome, and M. Gianella: Infrared laser-based sensing in medical applications, *Proc. SPIE* **7608**, 760808-1 – 760808-9 (2010).

[120] D. E. Ott: Laparoscopic Hypothermia, *J. Laparoendosc. Surg.* **1**, 127–31 (1991).

[121] T. Katoh and K. Ikeda: The Minimum Alveolar Concentration (MAC) of Sevoflurane in Humans, *Anesthesiology* **66**, 301–303 (1987).

[122] OSHA: http://www.osha.gov/dts/osta/anestheticgases/index.html.

[123] C. Keijzer, R. S. G. M. Perez, and J. J. de Lange: Compound A and Carbon Monoxide Production from Sevoflurane and Seven Different Types of Carbon Dioxide Absorbent in a Patient Model, *Acta Anaesthesiol. Scand.* **51**, 31–37 (2007).

[124] A. Brown, C. Canosamas, A. Par, J. Pierce, and R. Wayne: Tropospheric Lifetimes of Halogenated Anesthetics, *Nature* **341**, 635–637 (1989).

[125] E. Esper, T. E. Russell, B. Coy, B. E. Duke, M. H. Max, and J. A. Coil: Transperitoneal Absorption of Thermocautery-Induced Carbon-Monoxide Formation during Laparoscopic Cholecystectomy, *Surg. Endosc.* **4**, 333–335 (1994).

[126] C. Nezhat, D. S. Seidman, H. J. Vreman, D. K. Stevenson, F. Nezhat, and C. Nezhat: The Risk of Carbon Monoxide Poisoning after Prolonged Laparoscopic Surgery, *Obstet. Gynecol.* **88**, 771–774 (1996).

# Publications

## Book Chapters

H. Wächter, M. Gianella, and M.W. Sigrist
*Generation of Coherent Mid-Infrared Radiation by Difference Frequency Mixing*
in Springer Handbook of Laser and Optics, edited by F. Träger, 2nd edition, Springer, Berlin, Heidelberg (*in print*).

## Refereed Journals

M. Gianella, J. M. Rey, D. Hahnloser, and M. W. Sigrist
*$CO_2$ Laser-Photoacoustic Analysis of Smoke Emitted during Minimal Invasive Electro-Knife Surgery*
European Physical Journal – Special Topics **153**, 459–462 (2008).

M. Hübner, M. W. Sigrist, N. Demartines, M. Gianella, P. A. Clavien, and D. Hahnloser
*Gas Emission during Laparoscopic Colorectal Surgery Using a Bipolar Vessel Sealing Device: A Pilot Study on Four Patients*
Patient Safety in Surgery **2**, 22 (2008).

M. Gianella and M. W. Sigrist
*Improved Algorithm for Quantitative Analyses of Infrared Spectra of Multicomponent Gas Mixtures with Unknown Compositions*
Appl. Spectrosc. **63**, 338–343 (2009).

M. W. Sigrist, R. Bartlome, and M. Gianella
*Infrared laser-based sensing in medical applications*
in Quantum Sensing and Nanophotonic Devices VII
Proc. SPIE **7608**, 760808–1–760808–9 (2010), invited paper.

M. Gianella and M. W. Sigrist
*Infrared Spectroscopy on Smoke Produced by Cauterization of Animal Tissue*
Sensors **10**, 2694–2708 (2010).

M. Gianella and M. W. Sigrist
*Automated Broad Tuning of Difference Frequency Sources for Spectroscopic Studies*
Appl. Opt. **50**, A11–A19 (2011).

# Proceedings

D. Marinov, J. M. Rey, M. Gianella and M. W. Sigrist
*Human Breath Analysis Employing DFG Laser Spectroscopy*
Dig. CLEO/Europe 2005, Munich (D), June 12–17 (2005)
European Conf. Abstracts, Vol. 29B, paper CD-3-MON.

M. Gianella, J. M. Rey and M. W. Sigrist
*$CO_2$ laser-photoacoustic analysis of smoke emitted during electro-knife surgery*
Dig. 14th Int. Conf. on Photoacoustic and Photothermal Phenomena (ICPPP), Cairo (Egypt), January 6–9 (2007), p. 206.

M. Gianella and M. W. Sigrist
*Infrared laser spectroscopy on surgical smoke*
Dig. CLEO/Europe–EQEC 2009, Munich (D), June 15–19 (2009).

M. Gianella and M. W. Sigrist
*Laser spectroscopic studies on surgical smoke*
Dig. 7th Int. Conf. on Tunable Diode Laser Spectroscopy (TDLS-2009), Zermatt (CH), July 13–17 (2009), paper B13, p. 54.

M. W. Sigrist, R. Bartlome, and M. Gianella
*Medical applications of laser spectroscopic gas analyses*
Dig. 17th Int. Conf. on Advanced Laser Technologies (ALT09), Antalya (Turkey), Sep 26–Oct 1 (2009), p. 177.

M. W. Sigrist, R. Bartlome and M. Gianella
*Infrared laser-based sensing in medical applications*
Dig. OPTO/SPIE Photonics West, San Francisco (USA), Jan 23–28 (2010).

M. Gianella and M. W. Sigrist
*Application of Mid-IR Laser Spectroscopy for the Analysis of Surgical Smoke*
Dig. Laser Applications to Chemical, Security and Environmental Analysis (LAC-SEA), San Diego (USA), Jan 31–Feb 3 (2010).

# Contributions at Conferences

M. Gianella and M. W. Sigrist
*Infrared laser spectroscopy on surgical smoke*
CLEO/Europe 2009, Munich (D), June 15–19 (2009).

M. Gianella and M. W. Sigrist
*Laser spectroscopic studies on surgical smoke*
7th Int. Conf. on Tunable Diode Laser Spectroscopy (TDLS-2009), Zermatt (CH), July 13–17 (2009).

# Acknowledgments

I want to express my gratitude to Prof. Dr. Markus Sigrist for providing an agreeable research atmosphere within the group, and for always being available for discussions and advice.

My thanks go to Prof. Dr. Jérôme Faist and Prof. Dr. Andrew Orr-Ewing for reading my thesis and being my co-examiners.

I thank Dr. Julien Rey for many enlightening conversations, and my fellow doctoral students – Kerstin Hans and Jonas Kottmann – for helpful suggestions and a pleasant working environment.

I would like to thank the three semester students – Qiang Cui, Sabine Riedi and Jonas Alpstäg – who carried out several measurements.

Many thanks to PD Dr. Dieter Hahnloser of the University hospital Zurich for accepting an outsider in the operation room and for allowing me to collect samples.

I am also grateful to Dr. Georg Seyfang of ETH Zurich for helping with some of the measurements.

Finally, I want to thank my family for their support.

The financial support by the Swiss National Science Foundation and ETH Zurich is gratefully acknowledged.

Die VDM Verlagsservicegesellschaft sucht für wissenschaftliche Verlage abgeschlossene und herausragende

## Dissertationen, Habilitationen, Diplomarbeiten, Master Theses, Magisterarbeiten usw.

### für die kostenlose Publikation als Fachbuch.

Sie verfügen über eine Arbeit, die hohen inhaltlichen und formalen Ansprüchen genügt, und haben Interesse an einer honorarvergüteten Publikation?

Dann senden Sie bitte erste Informationen über sich und Ihre Arbeit per Email an *info@vdm-vsg.de*.

### Sie erhalten kurzfristig unser Feedback!

VDM Verlagsservicegesellschaft mbH
Dudweiler Landstr. 99           Telefon +49 681 3720 174
D - 66123 Saarbrücken           Fax     +49 681 3720 1749
### www.vdm-vsg.de

Die VDM Verlagsservicegesellschaft mbH vertritt

Printed by Books on Demand GmbH, Norderstedt / Germany